Occasional Paper Series 7

The Anti-Tactical Ballistic Missile Issue and European Security

Ian M. Cuthbertson

WESTVIEW PRESS * BOULDER, COLORADO

Institute for East-West Security Studies
New York 1988

The Institute for East-West Security Studies does not take or encourage specific policy positions. It is committed to encouraging and facilitating discussions of important issues of concern to East and West. The views expressed in this book do not necessarily reflect the opinions of the Board of Directors, the officers or the staff of the Institute.

This publication is made possible through the generous support of the Ford Foundation.

Distributed by Westview Press
Frederick A. Praeger, Publisher
5500 Central Avenue
Boulder, Colorado 80301

Library of Congress Cataloging-in-Publication Data

Cuthbertson, Ian M., 1957–
The anti-tactical ballistic missile issue and European security.

(Occasional paper series/Institute for East-West Security Studies; 7)
Bibliography: p.
1. Europe—National security. 2. Anti-tactical ballistic missiles—Europe. 3. Europe—Air defenses, Military. I. Title. II. Series: Occasional paper series (Institute for East-West Security Studies); 7.
UA646.C87 1988 355′.03304 88-12749

ISBN 0-913449-05-9 (IEWSS)
0-8133-0660-4 (Westview)

Printed in the United States of America

TABLE OF CONTENTS

FOREWORD

The feasibility and desirability of deploying anti-tactical ballistic missiles (ATBMs) has long been overshadowed by the wider debate on ballistic missile defense (BMD), in particular the Strategic Defense Initiative (SDI). This is unfortunate, as the issues involved in providing a defense against conventionally armed missiles are substantially different from those involved in countering strategic nuclear ballistic missiles.

Ian Cuthbertson's paper is part of a larger project the Institute has been conducting on the implications of new defensive systems for stability and security in Europe. It seeks to look beyond the narrow technical issues involved with anti-missile systems, and to place the whole issue within the wider political and military questions involved in any plan to develop and deploy ATBMs. Taking as a given the technological ability of both NATO and the Warsaw Treaty Organization (WTO) to build ATBMs, the paper seeks to explore the issues which ATBMs raise for NATO in terms of political and military priorities, alliance strategy, burden-sharing and arms control. Cuthbertson also gives some idea of what the nature of the debate within the WTO might be as it confronts some of the same military problems faced by NATO.

The paper concludes that while feasible, ATBMs are too expensive a solution to the possibility of conventionally armed tactical ballistic missiles (TBMs). The strain on defense budgets in both alliances has forced them to prioritize threats to their security. Conventional TBMs come too far behind the threat from aircraft to justify a dedicated response. Passive measures of defense, involving the duplication, hardening and dispersal of potential targets appear to offer a more cost-effective and multi-purposes response. Such an approach would help counter the threats from both conventional TBMs and aircraft, as well as providing some measure of protection from other threats, such as attacks by special forces troops.

The author, Ian M. Cuthbertson, was a Senior Research Officer in the Foreign and Commonwealth Office of the United Kingdom. He wrote this paper while a Resident Fellow at the Institute for East-West Security Studies in 1986–1987, where the paper was exposed to the critical review of an East-West team of foreign policy and security experts. The author wishes to express particular appreciation to the Resident Fellows of the Institute for 1986–1987, in particular Joachim Krause and Alain Carton, and to Institute staff members Dr. F. Stephen Larrabee, Dr. Allen Lynch and Peter B. Kaufman for their comments and handling of the manuscript from start to finish.

The Institute gratefully acknowledges the support of the Ford Foundation for providing the financial assistance for this study, its publication and dissemination. The views in this paper are solely those of its author and should not be ascribed to the funders, the Institute, or the government of the United Kingdom. The Institute for East-West Security Studies is pleased to sponsor its publication as a contribution to the debate in both East and West about this important issue of international security.

John Edwin Mroz
President
March 1988

1
Introduction

The debate which surrounds the Strategic Defense Initiative (SDI) research program and the changing strategic environment which it presages has also acted as a catalyst for a lively debate on the feasibility and desirability of anti-tactical ballistic missile defense (ATBM) against theater-range tactical ballistic missiles (TBM) targeted within Europe. Although the issue of ATBMs has in part become inexorably linked to the broader SDI debate, the problem of TBMs and the possibilities of a solution based on ATBMs predates by three years the revived debate surrounding strategic defense. Indeed, in many ways it is a serious mistake to mix the two debates together. This linkage has tended to divert political and public attention from the real issues surrounding ATBMs and instead has focused attention on defending against a nuclear ballistic missile attack.

Nuclear attacks, however, are not the essence of the ATBM issue. It is certainly not the problem with which both the North Atlantic Treaty Organization (NATO) and the Warsaw Treaty Organization (WTO) are grappling. The real debate within NATO and the WTO focuses on the possible threat to important military targets from increasingly accurate tactical ballistic missiles which could be armed in the near future with *conventional* warheads. The solution which is being sought to this problem is a way to protect, either by active or passive means, valuable infrastructure targets, such as air fields and C^3I installations. At no time has the European debate ever seriously considered, unlike the debate surrounding SDI and strategic ballistic missile defense (BMD), the possibility of a reliable and cost-effective defense against the nuclear TBM threat.

The December 1987 conclusion of the Intermediate-range

Nuclear Forces (INF) Treaty has done little to change the motivations and rationales which underpin the ATBM debate in Europe. Despite the putative withdrawal of a large number of missiles from the inventories of both alliance systems—missiles which might have had some capability as conventionally armed delivery vehicles—a large number of missiles remain which could fulfill a similar function. Calls have already been made for NATO to replace the withdrawn Pershing II and cruise nuclear missiles with conventionally armed sea-launched cruise missiles.[1] It is among such replacement systems, specifically the new generation of shorter-range (SR) INF now under development for both NATO and the WTO, that the future targets of any ATBM system lie.

With the possible exception of the Pershing II, none of the ballistic missiles eliminated by the INF Treaty had any realistic possibility of functioning effectively with a conventional warhead. Their accuracy was not sufficiently high to guarantee destruction of a target with the greatly reduced explosive payload which a conventional warhead offers compared to even the smallest nuclear warhead. However, the new generations of SRINF promise the type of accurate delivery which may make possible the use of conventional explosives for bombardment and interdiction of specific, militarily valuable targets up to ranges of 500 kilometers. Both NATO and the WTO continue to believe that this type of attack offers one of the most effective ways of crippling the war-fighting potential of its opponent. Both, for their own reasons, are increasingly unwilling to assign such tasks solely to nuclear weapons. As long as any ballistic missile remains in service which is capable of delivering either a nuclear weapon or a conventional warhead, then the value and cost-effectiveness of deploying defenses to meet the conventional threat will continue to be debated.

ATBM research thus seems set to continue regardless of whether there are further moves to reduce the number of nuclear weapons in Europe. Such research is considered as a useful hedge against decreasing ballistic missile circular error probables (CEPs) and possible breakout from the existing INF treaty or any future SRINF arms control treaty. The U.S.

---

1. *Financial Times,* February 10, 1988, p. 20.

military budget for fiscal year 1988-1989 includes an allocation of $280 million for the U.S. Army's anti-tactical missile program, which encompasses the study of active defense, passive measures and even counterforce options. This proposal represents a major increase over the $71 million allocated in 1987, and shows the seriousness which the United States places on the threat from TBMs.[2]

The serious threat which some American policy-makers see from the future developments in the field of conventional TBMs is most clearly stated in the 1988 report of the Commission on Integrated Long-Term Strategy, entitled *Discriminate Deterrence*. This report, drawn up by a small group of eminent academics and senior government officials, seeks to map out a U.S. military strategy in a world with much fewer nuclear weapons. The Commission's report states boldly that:

> . . . The issue of missile defense is raised urgently by the Soviet acquisition of ballistic missiles accurate enough to threaten many critical European targets with conventional warheads or chemical weapons. Among the targets under the gun: operating bases for tactical aircraft, command-and-control centers, and major supply centers.
>
> We need both active and passive defenses against the missiles. Passive measures should include hardening, dispersal, multiplication of critical facilities and stockpiling of rapid-repair equipment. The active defenses should probably include some directed at tactical ballistic missiles (ATBM systems), although the INF treaty should help reduce the threat from them. Assuming that we opt for ATBM, upgrading current air defenses would look like a useful first step, but only that. More capable "extended" air defense, now under study in NATO, would require advanced endo-atmospheric interceptors and probably airborne optical sensors. Components with these capabilities can be derived from existing technology for defense against strategic missiles.
>
> Ballistic missile defense is important in a broad range of situations. Active defenses with even moderate capabilities can contribute to deterring Soviet attacks on many different theater targets . . . . The evolutionary development of missile defenses should be continued. Improving missile and air

2. See U.S. Department of Defense, *Annual Report to the Congress: Fiscal Year 1988* (Washington, DC: U.S. Government Printing Office, 1987).

defense will eventually make possible a real reduction in the terrors of the nuclear age. It might also serve to moderate our concerns about the breakdown of an arms control agreement providing for deep cuts in offensive missiles.[3]

It is clear that the possibility of deploying accurate conventionally armed tactical ballistic missiles lies within the grasp of both NATO and the WTO. The technical capability to produce active defenses is also within the reach of both alliances. Both alliances are concerned with the ramifications of this debate. There are, therefore, two main questions which both alliances must address when looking at the issues raised by conventional TBMs. Firstly, what contribution can such weapons make towards improving the capabilities of their respective military forces for waging a conventional war in the European theater? Secondly, what can active defense for the prospective military targets of conventional TBMs contribute towards negating any advantages which such missiles may confer? For the West in particular, the issue focuses on what can be done to ensure that irreplaceable military assets are less vulnerable to a Soviet preemptive conventional ballistic missile strike aimed at seriously damaging, at least for a short period of time, NATO's defensive capabilities. For the WTO, the issue is how to avoid serious disruption of its capability to defend itself through offensive action in the European theater, and how to protect its forces from such emerging NATO concepts as follow-on-forces attack (FOFA).

This paper sets out to analyze the threat which both sides perceive from the deployment of increasingly accurate TBMs. The concern felt by both alliances is that accurate theater-range ballistic missiles, when coupled with the possibility of improved explosive technology, may give TBMs a significant conventional capability. For NATO, the problem revolves around the SS-23 and the new Soviet SRINF which may soon be introduced into Eastern Europe. It is believed that these weapons could pose a credible threat to NATO targets if they were to be armed with conventional warheads. The debate within NATO focuses on how accurate the existing generation of missiles actually is, and the realistic time scale for further

3. Fred C. Iklé and Albert Wohlstetter, *Discriminate Deterrence: Report of the Commission on Integrated Long-Term Strategy* (Washington, DC: U.S. Government Printing Office, January 1988), pp. 51–52.

improvements in the accuracy of both these and future generations of missiles and the types of conventional or chemical warheads they could be capable of carrying.

The debate also revolves around what proportion of the Soviet TBM force could be allocated to conventional attacks and what their potential targets may be. NATO has tended to focus on the threat to such fixed installations as air fields, C^3I installations, nuclear weapon storage and military equipment dumps. However, attention has also begun to focus on a possible Soviet threat to NATO reinforcements heading for the front, using TBMs programmed with real-time intelligence data on troop movements. The whole ATBM debate, therefore, revolves around what contribution active defense can make to the protection of a wide range of assets, and in what ways active defense can supplement or replace passive defenses.

The active defense debate, in turn, is carried out against the backdrop of SDI. When, after President Reagan's March 1983 SDI speech, NATO's political leaders began to look seriously at the issues involved in defending against TBMs, the idea of boost- and mid-phase interception was casually bandied around. However, as I noted at the outset, this debate was in many ways superimposed artificially on an existing discussion going on within NATO's military/technological communities on how to defend against TBMs using available technology. This original and narrower debate gained momentum within the alliance as it quickly came to be realized that the exotic approach to defense against TBMs, based on lasers or particle-beam weapons, was likely to be unrealizable for reasons of economy, allied to a lack of the necessary indigenous technological capability among NATO's European allies. The United States was quick to point out that it would not be responsible for paying for the deployment of a European ATBM system. Any SDI-derived technology incorporated into such a system would, therefore, have to be paid for from West European defense budgets. In addition, arms control concerns, focused on the technology-transfer provisions of the ABM Treaty, caused many European politicians to query the idea of using exotic technologies for TBM defense.

The growing realization of the problems involved in the SDI-type approach to ATBM encouraged the focus of the discussion to revert to where it had originally been when it was simply a military/technical debate between experts. Thoughts again turned to the possibility of preemption of Soviet TBM launch, as originally set out in a number of NATO air-defense studies in the early 1980s.[4] However, at this point the political dimensions of this debate reasserted themselves and it was quickly realized that whatever the technical and military appeal of preemption as a defense strategy, given NATO's political dynamics it was likely to be an impossible strategy to implement. Discussion within NATO then focused on the possibilities of some form of air-defense upgrade, with a number of missiles touted as possible candidates for dual ATBM/air-defense capability. The Patriot missile was considered by a number of defense analysts as a prime contender for this role.[5]

This paper will discuss some of the pros and cons of this approach to active defense. From a technical point of view, NATO is certainly capable of providing some of its air-defense missiles with a limited ATBM capability. It may even be desirable for NATO to follow this path. It has been suggested that one of the prime targets for any selective Soviet conventional TBM attack would be these air-defense missiles and their associated radar systems,[6] to make life easier for Soviet frontal aviation in NATO airspace. However, it is at this point that the ugly serpent of affordability raises its head. Beyond some marginal upgrade for self-defense for air-defense missiles, is ATBM a cost-effective option for NATO to follow? Given the ever-growing pressures on alliance defense budgets, and the long list of alternative priorities, it seems that the answer to this question may be no. In addition, there are also serious issues of compliance with the ABM Treaty which could arise from this form of active defense. The ATBM issue also helps to

---

4. AGARD, the advisory group for aerospace research and development, was set up to exchange scientific and terminal information between NATO states. It was never intended to be a decision-making body, although the depth and breath of knowledge which its participants bring to an issue gives considerable force to its recommendations.
5. See Ivo H. Daalder, "A Tactical Defense Initiative for Western Europe?" *Bulletin of the Atomic Scientists* 43, No. 4 (May 1987), pp. 34–39; also *The Economist*, January 12, 1985.
6. See Benoit Morel and Theodore A. Postol, "Non-Nuclear Soviet Tactical Ballistic Missiles: A Threat to NATO?" in Jurgen Altmann, et al., eds., *Anti-Tactical Missile Defenses and West European Security*, HSFK-Report 4/1987 (Frankfurt: Hessische-Stiftung Friedens- und Konfliktforschung, 1987), pp. 45–79.

crystallize a number of questions about the possibilities for further arms control in the area of both defensive and offensive missiles. What could be the impact on the ATBM debate of the successful implementation of a disarmament treaty, built on the success achieved during the INF negotiations, and intended to reduce or eliminate shorter-range TBMs from Europe? With all of these issues in mind, this paper will discuss whether there are incentives for both alliances to explore other approaches, apart from anti-missile systems, to dealing with a conventional TBM threat.

The other major approach to defense against TBMs embraces a number of approaches which have collectively been described as passive measures. This term covers a wide variety of activities, from major civil engineering projects including such things as pouring concrete to increase the number of runways and taxi-ways at air fields and harden aircraft shelters; important procurement decisions, such as initiating improvements in NATO's electronic countermeasures capability; down to small, simple steps, for example training more bomb-disposal personnel, to speed up the return of installations to full operational capability after an attack. The central question is what contribution can such measures make to TBM defense and what might be their usefulness against other threats to these installations, such as the air threat? This point is of major importance. The threat to NATO's military infrastructure from Soviet frontal aviation is widely recognized. Unlike conventional TBMs, it is not a gleam in a politician's or weapon designer's eye. It is, rather, a genuine threat which NATO planners have already been seriously questioning the capabilities of the alliance to adequately cope with. In discussing ATBMs, it is important to bear in mind that, regardless of how accurate future TBMs may be, or how powerful a conventional warhead they can deliver, they are unlikely to be viewed as more than a useful prelude or adjunct to the far more powerful and flexible strikes which the air forces of both alliances are designed to deliver. Both passive and active ATBM defenses are likely to have to contribute significantly to blunting the tactical air threat if they are to be judged cost-effective.

The cost-effectiveness of the passive approach will also be judged on the question of where the funds could be found to implement such measures. One of the particular appeals of this approach may be that the NATO infrastructure budget can be made to bear many of the costs of passive defense, thus sharing the economic burden more evenly throughout the alliance.

As previously mentioned, ATBM research and development is set to continue. Within NATO, there is a vocal constituency, which has grown up partly as a result of the SDI debate, to investigate and exploit the possibilities of active defense. But beyond a generally perceived need for low-level prophylactic R&D, there is an opposite constituency, especially among the military in Europe, also strong although less vocal, who do not favor ATBM deployment. They see no major threat in the foreseeable future from conventional TBMs and they would much prefer any available resources to be spent on what they see as far more pressing military needs, such as the tactical air threat. There are also strong counter-incentives on the part of West European politicians, who do not wish to see themselves forced into a position of being seen to be deploying SDI-derived or SDI-related technology. In this regard the start of the early-deployment debate in the United States had an obvious and rapid impact on the ATBM debate in Europe, in that it has taken much of the heat out of the European ATBM debate.

If much of the topicality has gone out of the current round of the discussions on ATBM now that the entire subject has been largely removed from its previously high political profile, it then remains to assess what the consequences are for the possibility of an ATBM deployment by NATO. What are the future implications now that the debate has reverted to its original military/technical level? What time scale do NATO's military decision-makers view as realistic for a more serious danger from conventional TBMs emerging, and what steps should the political leadership currently take in both alliances to prepare for this future threat?

2

The Strategic Environment

Doubts about Deterrence and Disarmament

For thirty years the security of Europe has rested, *de facto,* if not *de jure,* on the existence of a rough balance of nuclear capabilities between the two superpowers. Nuclear deterrence has thus been seen as the ultimate guarantee of peace in Europe. However, improved technology has or is about to make available to both alliance systems missiles whose accuracy makes counterforce targeting against land-based nuclear missiles a realistic option. For NATO this means that long-held fears of a disarming first-strike by the WTO, which date back to the late 1960s and the initial deployment by the Soviet Union of SS-9s, begin more and more to look like a practical proposition. Deterrence has become a prisoner and a victim of its own success. The very existence of the knowledge of how to create nuclear weapons makes total nuclear disarmament both undesirable and impossible even as a long-term goal. At the same time, the approach being adopted by the nuclear-weapon states, which is based on programs aimed at their constant improvement, seems to lessen rather than increase the security of land-based nuclear assets.

It is ironic that the pace of technological development and the possibilities which it offers threaten to undermine the rough stability on which nuclear deterrence has rested for so long. It is this technology which has made possible the vast proliferation of both the numbers and capabilities of missiles and nuclear warheads. It is this proliferation, and the increased uncertainty which technological developments have brought to the nuclear balance, which have given rise to the current debate on the effectiveness and usefulness of deterrence. These same developments have also raised serious doubts among the general public, in both East and West, over the desirability of a policy which ultimately rests on a mutual suicide pact.

In the West, there no longer exists a public and political consensus on the usefulness of deterrence. Large public demonstrations in Western Europe against the stationing of cruise and Pershing II missiles reflect a perception that the nuclear arms race is out of control. This perception in Western public opinion has coincided with the advent of a dynamic and photogenic new Soviet leader. The Soviet Union and its allies have sought to exploit the concerns felt by Western publics, and in doing so they have used an innovative and flexible style. The new Soviet style depends on superficially attractive proposals on both nuclear and conventional issues which are sweeping in their intent and designed to catch the interests of the Western media and, through them, the wider public. This approach, which depends far more on sophisticated public relations and less on old-fashioned propaganda and rhetoric, has also faced Western leaders with a challenge of a kind which previous Soviet leaders had conspicuously failed to develop. The Soviets have also been careful to ensure that each proposal contains a kernel of a negotiable proposal. This approach, allied to a wholly new and more forthcoming attitude towards the question of intrusive verification, has made it difficult for Western governments, especially in Europe, to be as dismissive of new WTO proposals as they have been in the past.

Added to this, the advent of SDI and the initiation of a widespread debate on the offense/defense relationship further confuses both the criteria for determining strategic stability and ultimately the esoteric calculations of what could constitute a successful first strike. It is on such abstract calculations of both conventional and nuclear capabilities by military experts on both sides that the success of a defense based on 'a seamless web of nuclear deterrence' ultimately depends.

The major characteristic of recent thought and discussions on the possible creation of effective ATBM systems also raises serious questions about the desirability of eliminating, through disarmament, the possibility of effective use of nuclear-armed missiles from the European theater. If there are no nuclear ballistic missile systems in Europe and both sides can counter or absorb a conventional strike by ballistic missiles, what then remains of deterrence? Both sides would return to a situation where a failure of conventional forces on the European mainland could be reacted to only by the use of strategic systems, with all the implications which such an approach entails. There is thus a strong argument for retaining nuclear weapons as a necessary and desirable tool of political and military decision-making, whose possible use is embodied in the escalatory ladder of NATO's doctrine of flexible response. Whatever the merits of conventionally armed ballistic missiles and ATBM, nuclear weapons remain the ultimate guarantee that NATO's European partners can continue to rely on deterrence and not defense for their security.

The debate within NATO on ATBMs is one which sets out to define, in classic military/technical terms, an effective response to an emerging threat, in this case identified as conventionally armed TBMs which may in the future be deployed by the Soviet Union. At this stage it should be pointed out that, due to the lack of reliable information on WTO activities, this paper tends to focus on NATO's threat perceptions and internal debate, along with the problems which have flowed from this. However, the prospect of conventionally armed ballistic missiles is one which is of equal concern to the countries of the Warsaw Treaty Organization. Their threat perceptions cannot be radically different from those held by NATO. They face similar types of ballistic missiles and have a strong interest in protecting similar assets. What are the implications for WTO military strategy if account needs to be taken of NATO conventional TBM capability? What interests might the WTO have in developing its own ATBM? Given the widespread discussion in the Soviet Union of the dangers inherent in NATO adopting a strategy based on follow-on-forces attack (FOFA), not to mention air-land battle and the Rogers plan, there is no reason to believe that a debate is not also being carried out within the WTO on how to deal with an emerging threat from NATO of conventionally armed TBMs.

The imperatives and structures of this debate may be different. However, the existence of such systems as SA-10 and SA-12 and their testing and deployment as anti-ballistic missiles show that the Soviet Union may also be concerned by the threat which may be posed by conventionally armed tactical missiles. Indeed, threat perception in the East may be stronger than among NATO countries as the Soviet Union must be well aware that, given the relative crudeness of its own technology, it is less capable than the West of dealing with a nuclear-armed TBM. Therefore such ATBMs as it deploys may only be intended to function effectively against conventionally armed missiles.

A strong incentive for the Soviet Union in pursuing ATBM against the conventional TBM threat is the relatively few numbers of missiles which NATO is likely to have available for use in this way. NATO's traditional reliance on its judgment of what constitutes minimum deterrence has led to small numbers of deployed missiles, whose unit cost is very high. Even if a shift is made to conventional TBMs, it remains unlikely, for purely financial reasons, that the actual numbers of missiles deployed will increase much. With relatively few NATO missiles stretched over a large number of targets, it becomes worthwhile for the Soviets to attempt interception in the hope that only one or two NATO missiles can be devoted to each target. In this situation any interception offers the possibility of a 50-percent or 100-percent success rate for the defense. Thus while the Soviet Union may be preparing to counter a threat which does not yet exist, it could be doing so in the belief that it is a threat to which it should and can have an adequate response.

■ *Shifts in Strategy*

The ATBM debate among NATO countries arose because two major shifts are taking place in American strategy, both of which affect stability in Europe. The first of these shifts is embodied in the issues surrounding SDI. Although SDI is not the sole initiator, the political and technical debate surrounding it has acted as a major catalyst for a number of discussions designed to explore subsidiary and related issues. Originally focused on such ideas as an 'astrodome' over either the United States, the Soviet Union, or both, the SDI debate quickly focused on discussion of the possibilities of a move away from deterrence based on pure offense to a situation where there is a mixture, a relationship, between defense and offense. This topic was originally reflected in the purely military/technological discussion which had been taking place in NATO, and also Soviet, defense thinking as new weapons systems appeared to offer new possibilities. It was President Reagan's SDI speech which brought this debate to public attention and placed it on the political agenda of NATO. The second strand of thought originally dealt with more overtly political considerations, concerning moves towards the no-early-use of nuclear weapons. This shift represented a growing political and public disenchantment with the high numbers of nuclear weapons currently deployed both at the strategic and the theater level. Although there were different motivations for these two strands of the debate, the consideration given by politicians to an offense/defense relationship, including the serious discussions which surrounded the possibility for active defense of military targets against both nuclear and conventional ballistic missile threats, has been a response as much to the disenchantment felt by many over the growth in the numbers and capabilities of nuclear weapons as to the technological possibilities offered by new and, in many cases, revolutionary technology.

It needs to be emphasized, however, that the serious thinking which is now being done about defense within NATO represents, in its political manifestations, a major shift for the alliance. Previously, all partners had been content to rely on nuclear deterrence based on offensive weapons. Indeed, for many of the West European partners this continues to be the preference. It is the United States which is the driving force behind the debate. For the Soviet Union, the situation is somewhat different. Defense has always been a high Soviet priority, and BMD simply represents an extension of the concept into new areas. In the past, the Soviet Union's East European allies have appeared content with this approach. However, the wider political and military implications of a strategy which is based not on deterrence but rather on a more complicated formula including large-scale active defense may encourage some reexamination by East Europeans of the apparently largely unquestioned Soviet approach to these issues.

The motivation of calls on the part of some Western politicians and academic experts for NATO to implement a policy involving the no-early-use, or even no-first-use of nuclear weapons, arises from political considerations and not from any changes in the military situation facing Western Europe. It is public opinion in Western Europe that finds it increasingly unacceptable that NATO's defense doctrine relies on the early use of nuclear weapons were its conventional forces to crumble in the face of a Warsaw Pact offensive in Western Europe. Although NATO may gain some military advantages from the move away from nuclear weapons, in the shape of enhanced capabilities for its conventional forces, this essentially represents a bonus to what remains a largely myopic political imperative. For the Warsaw Treaty countries the conventionalization of the European theater has long been a priority objective.[7] It makes the WTO's current strategy, based on defense by offense, much less risky in terms of the possible use of nuclear weapons by NATO to defeat a conventional attack, be that use aimed at WTO troop concentrations in Eastern Europe and the Western military districts of the Soviet Union, or even if the nuclear response from NATO was to be targeted on command and control installations deep in Soviet territory. It was against this background that questions about where ATBM fitted into both the new discussion about the offense/defense relationship and the conventionalization and denuclearization of the European theater began to be raised.

■ *New Threats*

The current debate on the possible deployment of ATBMs in Europe has grown out of the perception among European nations that a new qualitative threat to their security has emerged. This threat takes the form of a generation of theater and tactical ballistic missiles whose circular error probable (CEP), or accuracy, is claimed to be very low, in the range of tens or hundred of meters. If such ballistic missiles were to be armed with conventional warheads, they would pose a significant new threat to militarily important targets in Western Europe by making possible their destruction without the use of nuclear weapons. For NATO in the late 1970s, the threat was initially identified as coming from a new Soviet theater-range missile, the SS-20. This was seen as the first Soviet intermediate-range ballistic missile (IRBM) with the capability to accurately strike targets in all of Western Europe

7. See Stephen M. Meyer, *Soviet Theater Nuclear Forces* (Parts I and II), *Adelphi Papers* 187 and 188 (London: The International Institute for Strategic Studies, 1984).

from the Soviet Union. It was later joined by new shorter-range INF (SRINF), in the shape of SS-21s, -22s and -23s, some of which have been stationed in Czechoslovakia and East Germany since 1983. These missiles are currently deployed along with a variety of battlefield ballistic missile systems, mainly SCUD Bs and a residual number of FROGs, which can be armed with either conventional, chemical or nuclear warheads, and which the Soviet Union has long had available in Eastern Europe. There is no evidence yet to suggest that the Soviet Union has begun to test any of its modern ballistic missiles, namely the SS-20, -22 or -23, with conventional warheads. This being the case, discussion about the need for ATBM remains largely theoretical for the time being.

It is reasonable to assume that NATO's threat perceptions are mirrored by concerns felt among Warsaw Pact countries. The introduction of the new generation of NATO theater nuclear ballistic systems, Pershing II, and the proposed French Hades, as well as ground- and sea-launched cruise missiles, pose a similar threat to the Warsaw Pact's infrastructure in Eastern Europe and the western Soviet Union. These missiles add significantly to the capability which NATO countries have deployed in the past, based either on ballistic missiles such as Pershing I and Lance or manned aircraft. Indeed, although much of the literature from the Soviet Union has dwelt on the nuclear threat posed by Pershing IIs for countervalue strikes on cities such as Moscow, an equally significant concern may focus on the threat which is perceived to exist from these missiles to the Warsaw Pact's defense capabilities.[8] This threat revolves around the new vulnerability of the Soviet command and control installations which cluster around Moscow—vulnerability not to selective nuclear strikes, but to a large-scale conventional attack.

Obviously, there arise serious military implications from the availability of long-range ballistic missiles which could be capable of rapidly destroying C^3I installations, air fields, nuclear weapon storage sites, ports and storage dumps without using nuclear warheads. Equally important, however, are the political implications of a large variety of ballistic missile systems deployed throughout Europe whose use is no longer

8. "U.S. Missiles 'Crippling Potential' to Security," *FBIS-USSR*, May 12, 1983, pp. CC1-CC3.

constrained by nuclear deterrence but which are viewed instead as militarily useful and usable weapons.

As an obvious example, this would have implications throughout the spectrum of launch scenarios. At present, early-warning launch-detection systems are incapable of distinguishing between a nuclear or conventionally armed ballistic missile. There is no available technology or any in prospect which would enable an early-warning system to verify the nature of a missile until impact had taken place upon a target, and even then low-yield nuclear weapons or the use of fuel-air explosives could still make it difficult to rapidly and accurately identify what type of attack was underway. Such a situation could lead to a launch-on-warning strategy, where a nuclear response would be implemented to what was, in fact, a conventional attack.

■ *Implications for Arms Control*

There are also serious implications concerning the development of conventional TBMs for the future of the arms control process in Europe. The debate over the American proposals put forward in October 1986 in Reykjavik for the elimination of all nuclear-armed offensive ballistic missiles has hardly even begun to delve into a discussion of what the implications are if such an agreement allowed for the retention of conventionally armed ballistic missiles. If, and it remains an unlikely if, the verifiable elimination of ballistic nuclear weapons were to be achieved, the role of conventionally armed ballistic missiles could quickly grow in importance. Such missiles promise the capability of the type of rapid strike, based on real-time intelligence, which cruise missiles and manned bombers cannot offer.

It therefore seems possible that even if a further agreement can be reached for the elimination of all nuclear-armed ballistic missiles in Europe, following on from the INF Treaty and embracing SRINF, there would still be considerable pressure from the military establishments of the two major powers, as well as from those of their major allies, for the retention of some ballistic missile capability. Realistically, it needs to be recognized that such revolutionary steps as the elimination of nuclear missiles must inevitably involve a dramatic reassessment of military force structures and doctrines by both alliance systems. In this situation conventionally armed ballistic missiles could be a vital component of both an offensive posture, aimed at a rapid advance through the enemy's territory, as well as a successful defensive posture designed to quickly blunt the cutting edge of any attack and also destroy the second echelon of an offensive. Such possibilities inevitably lead to the investigation of what steps can be taken to counteract or negate this potential.

A practicable defensive system against nuclear-armed missiles does not seem conceivable for Europe. The inevitable leakage which any ATBM system would suffer from, and the density of targets in both Eastern and Western Europe, means that the destruction wrought by even a partially blunted nuclear strike would be sufficient to totally destroy European countries as functioning societies. The chemical weapons (CW) threat posed by TBMs falls into much the same category as the threat posed by nuclear-armed missiles. While there may be some marginal utility in degrading a ballistic-missile-borne CW attack by achieving the destruction of a proportion of incoming missiles, the objective situation remains that from a military point of view much the same level of nuclear, biological and chemical (NBC) precautions have to be taken against such an attack whether one or 100 CW-armed missiles were to get through. It is a sad but true fact that little attention is devoted to the possibilities and benefits of defending civilians from such an attack. It therefore seems that whatever the effectiveness of active defenses, most of the onus would still be thrown onto passive CW defense measures for the protection of military targets. In addition, the relatively small sums now spent on CW defense would be likely to further suffer should funds be diverted into active defense. ATBMs therefore do not appear a very efficient way, especially for NATO, of going about meeting the CW threat, even should the threat emanate from TBMs. ATBM, therefore, only appears to become a viable option if it is considered in conjunction with attempts to blunt a conventional ballistic missile attack. Here, the degradation of Soviet offensive missile capabilities offer tangible military rewards in terms of protecting NATO installations. Similarly, for the Warsaw Pact, ATBM offers the possibility of protecting crucial command and control centers, as well as air fields, depot facilities and possibly even the critical choke points for reinforcing units in the field.

■ *ATBM Options*

While a new and dedicated ATBM system may offer one solution to these problems, it is also necessary to explore a whole range of other options with regard to defense against ballistic missiles armed with conventional warheads. A great deal of discussion has already taken place on what form of ATBM system would be most suitable for Western Europe. Proposals range from an upgrading of the current defense capabilities, based on existing radars and a new generation of Patriot air-defense missiles, through a new generation of ground-based missiles or directed-energy weapons dependent upon space-based tracking systems, up to the full panoply of an SDI-type system with both space-based target-acquisition and kill mechanisms for boost-phase and mid-course interception.

Although the technical possibilities of all of these options has excited wide discussion, the one certain fact concerning the available options for active defense is that they all would be extremely expensive. In the case of NATO, such deployment would require alliance-wide procurement decisions, similar to those which took place when NATO decided in the early 1980s to procure and deploy AWACs. The Byzantine negotiations which surrounded this procurement decision, and the fact that the U.K. in the end still opted out of the overall NATO decision, gives the flavor of what can be expected to occur as NATO moves towards a decision on whether ATBM is a viable option, and what form of defensive system promises to be the most suitable, affordable and achievable. Given the inevitably heated burden-sharing discussions which would take place both inside Western Europe and across the Atlantic, the possibilities for divisive conflicts within NATO are obvious.

■ *Non-ATBM Options*

With this potential friction within the NATO alliance in mind, it would obviously be worthwhile to also explore if there exist other alternatives to developing an active ATBM system for the defense of West European targets. These alternatives will have to meet Paul Nitze's criteria, set out in February 1985, of affordability, survivability and cost-effectiveness at the margin (relative to the effort and expense involved in procuring large numbers of new offensive weapons with the intention of swamping the defensive system).

Although Nitze's criteria were developed for SDI, they will inevitably also be used to judge what constitute an effective ATBM system. Passive measures of defense against ballistic missiles will be similarly judged. They must also be affordable, survivable and cost-effective at the margin. Would such alternatives be cheaper, more reliable and provide the same degree of security for major targets? Are the options of hardening, mobility and duplication feasible, and are they more or less effective than the variety of systems which may be possible for active defense? Similar judgments must apply to any decision by the WTO to deploy active defenses in Eastern Europe, but so far there has been no evidence to suggest that the WTO is engaged in the type of debate being carried on in Western Europe.[9]

There is also the consideration of what the consequences of ATBM procurement by either alliance would be for existing or future arms control agreements. The whole ATBM debate takes place with one eye on the ABM Treaty. The impact of this treaty on the type of ATBM system which NATO as an alliance might feel able to consider would be considerable. It is open to serious question if any of NATO's European allies would be prepared to consider procurement of a missile system which might appear to be in breach of the treaty, either in terms of its physical characteristics or in terms of the type of technological transfers to Western Europe which might be required from the United States to make the ATBM system a realistic possibility. For the Soviet Union, its close attachment to the restrictive interpretation of the ABM Treaty may make it unwilling to engage in any deployments which could credibly be used by the United States to advance further charges of non-compliance with the provisions of the treaty.

---

9. That the Soviets are aware of all these issues and are exploring them is shown by the attention given to NATO's ATBM and European Defense Initiative (EDI) debates by Soviet writers. See, for example, V. Beletsky, "What Lies Behind the 'European Defense Initiative' Project," *International Affairs* (Moscow) 6 (June 1986), pp. 49–55.

3

NATO Politics and the ATBM Debate

It would be a serious mistake to try and treat the issues surrounding the emerging threat to NATO assets from Soviet ballistic missiles and the questions which surround the desirability of ATBM as a unique or distinct element in the alliance debate on emerging problems which will face Western Europe in a few years. The whole ATBM issue is submerged in a much wider political debate on the future nature of NATO defense policy, issues of burden-sharing, resource management, weapons and equipment procurements, and especially multilateral projects and alliance arms control policy towards the Soviet Union.

Defense Policy

The most obvious solution to the ATBM issue is an arms control agreement which eliminates all TBMs from the European theater and prohibits not only nuclear-armed missiles but also any future plans to deploy conventionally armed missiles. However, the refusal of some of the NATO European allies to countenance what they view as the creeping denuclearization of Europe and their strong desire to retain a ballistic-missile component to their defense, including the possibility of conventionally armed missiles, makes this approach unrealistic. The political importance which NATO's European allies place on retaining TBMs in Europe would not even begin to be offset by any advantage gained from the removal of Soviet TBMs and nascent ATBM systems such as the SA-12 Giants or Gladiators. If the prospect of the elimination of Soviet TBMs does not excite the support of NATO's European allies, then throwing onto the scales the removal of the possibility of the Soviet Union's deploying ATBMs is unlikely to tip the balance.

Although Pershing II is about to be consigned to oblivion, the possibilities offered by its shorter-range successor systems, even in the conventional TBM mode, are still too little explored or understood for NATO to be anxious to foreclose that option. The Iklé report on *Discriminate Deterrence,* which sets out to map a long-term strategy for the United States, is clear on this point. The report, which throughout takes a military force planner's approach to future strategy and procurement, with a consequent overemphasis on narrow military-technical concerns, details the capabilities of these missiles:

> By the standards of a decade ago, the accuracies [of ballistic missiles] are extraordinary. Current technology makes it possible to attack fixed targets at any range with accuracies within one to three meters. These accuracies and modern munitions give us a high probability of destroying a wide variety of point and area targets with one or a few shots without using nuclear warheads. They make practical attacks on heavily defended military targets deep in enemy territory. Airfields well inside the Soviet Union could be put out of commission with warheads designed to attack infrastructure (fuel and maintenance facilities, say) and command-and-control facilities. Bridges, surface-to-air missile sites, intelligence facilities, rail lines, electric generating plants, petroleum refineries—all are suddenly much more vulnerable in the emerging age of smart munition.[10]

Notwithstanding the U.S. emphasis on capabilities, the importance which some of the NATO European allies—in particular France and the United Kingdom—place on the continued deployment of some form of nuclear TBMs makes all thoughts of the total elimination of these missiles and of ATBMs problematic. Clear evidence of this lies in the extensive debate now going on within NATO on the possibilities of the so-called triple-zero solution. While the FRG presses for negotiations with the Soviet Union aimed at the elimination of nuclear-armed TBMs from Europe, the United States, the U.K. and, to a lesser extent, France are pressing heavily for rapid

10. Iklé and Wohlstetter, *Discriminate Deterrence,* p. 50.

implementation of the 1983 NATO decision to modernize battlefield nuclear weapons.[11] With concerns over denuclearization countered by West German concerns about singularization (being the sole European target of nuclear weapons), no quick resolution is likely. This being the case, the status quo, in the form of deployment of nuclear TBMs, is set to continue.

It is, of course, possible to postulate radical changes in the military postures and doctrines of both alliances. Indeed, serious discussion has begun within NATO and the WTO of the possibilities for drastic realignment of doctrines, capabilities and hence perceived intentions. However, the military planning of either alliance cannot be based on postulated outcomes of putative negotiations on the future non-offensive doctrine or reduced military capabilities of an existing adversary. It therefore seems inevitable that the ATBM debate will continue even if its focus is narrowed down, from the Pershing II and SS-20 threat, with cruise missiles, SS-22 and -23s no longer in contention, but with discussion instead focusing on such battlefield systems as SS-21s, SCUDs and Lance. The critical point is that NATO's ATBM debate is here to stay and what is true for NATO applies equally to the WTO.

Studied in detail, the ATBM debate for NATO is not a coherent exchange of views on the emerging Soviet threat, nor is it a clear-sighted search for an antidote to a problem upon which all of NATO's diverse membership can agree. It is rather a disjointed discussion of different threat perceptions, based on often conflicting national intelligence estimates of Soviet capabilities, coupled with a search for the lowest common denominator among NATO's membership for feasible and affordable solutions to what appears to many as a chimera. As might be supposed, each country brings to the debate its own prejudices both in terms of perceptions of the threat and also what the preferred NATO response should be. This section of the paper sets out to discuss what shapes the attitudes and approaches of some of the major players in this game, and it will attempt to assess what impact competing national priorities could have on the final agreed NATO approach to any conventional TBM threat.

---

11. For U.S. Secretary of Defense Frank Carlucci's thoughts on NATO's tactical ballistic missiles, see *The New York Times*, February 8, 1988, p. A3.

United Kingdom

When it comes to assessing the feasibility and desirability of procuring and deploying ATBM systems, the United Kingdom is forced to judge the case not only on its own merits but also within the framework of its overall defense strategy and priorities. Whatever the ultimate utility and desirability of ATBMs, it is certain that this will not be the deciding factor in the decision of the U.K. to participate in such a program.

The overall British defense situation is largely governed by the four roles which the U.K. is assigned within NATO and the fiscal restraints which successive British governments have worked within in carrying out these tasks. The four main roles of the U.K. in NATO consist of the provision of nuclear forces, including the maintenance of an independent strategic nuclear deterrent; the defense of the United Kingdom itself; the provision of the land and air forces in Europe and naval forces in the eastern Atlantic and Channel areas; and contributing to forward defense in the Norwegian Sea. In turn, the equipment debate now going on within the United Kingdom is conditioned in part by decisions on where the unique priorities for the defense of the U.K. should lie, what overall strategy the United Kingdom should be pursuing within NATO for the collective defense of the alliance and the type of equipment which is necessary and affordable to achieve these objectives. Such considerations have complicated the debate which has surrounded the decision to acquire third-generation SLBMs for the U.K.'s strategic nuclear force, in the form of Trident D-5s, as well as such decisions as the building of the next generation of advanced fighter aircraft and procurement of AWACs.

Much of this debate has taken place in a charged political atmosphere which has seen, in the past eight years, the disintegration of the long-standing political consensus which has surrounded the major planks of British defense policy for the past thirty years. The two major British political parties now stand diametrically opposed on the need for an independent nuclear deterrent, with the Conservative Party supporting the modernization of Britain's independent nuclear deterrent and the Labour Party advocating unilateral nuclear disarmament. As a result, they are also far apart on what should be the future shape of NATO's strategy for the defense of Western Europe.

The most pressing limitation on the ability of the United Kingdom to fulfill adequately all of the commitments which it has traditionally undertaken is resource constraints, allied to some reluctance on the part of the British government to relinquish any of the tasks fulfilled in NATO by the British armed forces. After a number of years of real growth in the defense budgets, it is now the government's intention that the military budget should not increase in real terms. It also seems unlikely given the modest forecast of the rate of growth in the overall British economy that further resources are likely to be available in the near term. The dilemma for British policy-makers is that this freeze on real increases in the amount of resources allocated to defense coincides with the period of maximum bulge in the cost of procuring the Trident missile system. These costs come at a time when there is also a pressing need to improve the equipment of the British Army of the Rhine (BAOR) and the tactical airforce in Germany, along with those units assigned to the defense of the U.K. homeland.

In addition, the U.K. presently heads the league of spenders on military research and development among European NATO countries. Fifty-five percent of all British government-sponsored R&D goes to military projects. This is a much higher percentage than any of the U.K.'s NATO partners and the figure is highlighted by the fact that very little privately sponsored military R&D takes place within British industry. This is not the case in Germany or France, where a considerable amount of privately funded military R&D does take place. The British government is therefore likely to be very reluctant to produce further cash to participate in any R&D concerned with developing a new ATBM system. As one analyst has put it: "Many [British] government scientists, while recognising that some BMD was possible, believed that it would be very expensive and that effective offensive countermeasures would always be cheaper. Given the cost even of Patriot, few in the United Kingdom were impressed with the idea that a European effort at BMD/improved air defence might be worthwhile."[12]

In addition to the cost constraints, there is a more basic point that it is uncertain whether scientific and technical expertise could be found for such a project, given the fact that it

---

12. Trevor Taylor, "SDI—The British Response," in Hans-Guenter Brauch, ed., *Star Wars and European Defense* (London: Macmillan, 1987), p. 146.

is likely to be a major technological endeavor and that such manpower is already heavily engaged in other vital defense-related projects. Concern has already been expressed that the enthusiasm which British companies have shown in becoming involved in American-funded SDI research has already drawn irreplaceable technical and manpower resources out of indigenous military and civilian R&D. There also exists a strong fear that the United Kingdom will be denied access to the fruits of this research, both direct military technology and any civilian spinoffs.[13]

In addition to the debate going on within the defense community over resource allocation and the difficulty of slotting ATBM into existing priorities, there is also a wider debate which is taking place within the United Kingdom which has serious repercussions for the whole future of British defense policy. It is difficult to say where the resources can come from for ATBM, given the priority assigned to other programs. If, however, there were to be a change of government in four years' time and the main opposition party, Labour, were to take power, then there is little doubt that the Trident program would be cancelled and resources might become available both in terms of money and manpower for other projects. The Labour Party leadership has laid great emphasis on the need to improve the conventional defense capabilities of NATO, an approach intended to make their policy of British unilateral nuclear disarmament more politically credible, both to the British population and to Britain's NATO partners.

The implication of this approach is that the resources for this improvement in the U.K.'s conventional capabilities would come from the resources released by the abandonment of Trident and of the U.K.'s other nuclear tasks. However, the statements which the Labour Party has made on this issue suggest that such resources would be spent on other concerns, such as improving Britain's air-defense capability or the ability of BAOR to withstand a WTO attack by improving its conventional equipment. There has never been any suggestion that such resources could be devoted to any form of ATBM research or deployment. Due to the close linkages with SDI which the ATBM debate has established in both political and public

---

13. Paul Walton, "Facing the Consequences," *Informatics*, March 1986.

consciousness, such systems are unlikely to find favor with a substantial part of the Labour Party.

■ *Federal Republic of Germany*

In many ways the British case is the most extreme example of the difficulties facing a NATO country both in terms of resource constraints and also with regard to the collapse of the political consensus on defense issues. However, it must be realized that other countries face similar problems. The Federal Republic of Germany also has real constraints on the amount of resources it can devote to its military budget. The cycles of force modernization mean that a considerable amount of Bundeswehr equipment will be due to be replaced at the same time as money will have to be spent on ATBM development and deployment. This problem is exacerbated by the fact that the Bundeswehr is facing a critical manpower shortage which will see it short of 100,000 conscripts by 1990. Even the lengthening of the conscription period will not solve this problem of shortfall. If the Bundeswehr is unable to take in the 500,000 recruits it needs annually, then there is likely to be increased pressure for improving the equipment and firepower of Bundeswehr units to offset the numerical disadvantage. For a conscript army, even one as generous to its soldiers as the Bundeswehr, new equipment is always more expensive than soldiers.

In addition, the FRG suffers from a number of the same political problems which exist in the U.K. Within the West German government practical considerations seem to have triumphed over the initial enthusiasms of a number of senior West German politicians.[14] A large section of the Social Democratic Party (SPD), the main opposition party, has been very critical of the whole concept of ATBM, because of its close association with SDI. A number of prominent party members have expressed their outright opposition to the concept.[15] The prominence which was at one time accorded to the ATBM debate seems now to have all but disappeared in a welter of more immediate concerns. Once again, there does not seem to

---

14. See Manfred Woerner, "A Missile Defense for NATO Europe," *Strategic Review* 16, No. 1 (Winter 1986), pp. 13–20.
15. Hans-Guenter Brauch, "SDI—The Political Debate in the Federal Republic of Germany," in Brauch, ed., *Star Wars and European Defense*.

exist, at either end of the political spectrum, the will to embark on a development program, aimed at extensive ATBM deployments, when there are so many other more immediate calls on shrinking military budgets.[16]

France

While France suffers from many of the same fiscal constraints which exist in other West European countries, it alone of the major West European powers has retained its political consensus around defense issues. On the fiscal front, a significant French effort is at present concentrated on revamping and expanding the French nuclear deterrent. This program calls for the replacement of existing submarine hulls by new vessels along with an expansion in the numbers and capabilities of the SLBM force. The aircraft and land-based missile force are being modernized, as are the tactical nuclear weapons of the French army, with the replacement of Pluton missiles with Hades missiles. This concentration on nuclear forces has had some adverse affect on the ability of France to modernize its conventional forces. The army and the air force in particular have suffered, not so much in terms of the availability of new equipment, but rather in the quantities which can be procured. However, the restructuring of the French army and the decision to procure Hades shows that France is more committed than ever to the defense of Western Europe and in particular West Germany.

As stated previously, the consensus within the French political elite that nuclear weapons remain a desirable option has not been shattered in the way it has in other major West European countries. Similarly, the initial reaction to the American proposals on SDI was universally hostile. Only the realization that the West Europeans in general, and France in particular, could not afford to exclude themselves from the technological possibilities offered by SDI research made France reconsider the desirability of SDI-type research. However, this rationalization went hand-in-hand with a strong suspicion that the United States intended merely to milk the best of European talent and saddle Europe with the role of subcontractors. This was not a role which the French government was prepared to

16. Michael Feazel, "German Study Encourages Development of Anti-tactical Ballistic Missiles," *Aviation Week & Space Technology*, July 7, 1986.

consider. It therefore designed the Eureka high technology cooperation project as a civilian alternative to SDI military research. It was intended to provide the same kind of spark to technological innovation which the military side of SDI research was designed to impart to the U.S. economy.

The original reaction in France to the emerging ATBM debate in Western Europe was to lump it with the arguments and concerns surrounding SDI research.[17] On closer inspection of the ATBM issue, however, the French government identified a possible use for active defense to protect vulnerable nuclear missile and aircraft assets on the Plateau d'Albion. This realization led France to reconsider its position and to investigate what type of ATBM system might be feasible for the defense of military targets in Western Europe. France thus began to discuss with West Germany the possibilities of a European Defense Initiative (EDI), describing its approach as being aimed at a program which would carry out research into the possibilities for a European defensive system built and paid for by European nations although perhaps utilizing some American technology.[18] There were good fiscal reasons why the French felt that a partnership with West Germany was necessary in order to produce both the financial resources and the resources in trained manpower necessary to develop this type of system. However, it was also clear that the French government held out the bait of an ATBM project to try and forestall West German participation in SDI research. Thus, the German decision to participate in SDI research effectively buried what had already come to be seen as an overly grandiose approach to ATBMs.

There seems to be little questioning in France of the desirability of ATBM as a technical concept, rather only a problem with how to achieve what would seem to be a desirable military option. In developing this revised attitude to BMD, the French have been quick to take into account one of the major criticisms leveled at SDI, that it was overly ambitious and too expensive. The possibilities for using proven technologies, or technology which can be developed from existing systems plus the need to keep down costs, have come to

---

17. Alain Carton, "French Political Reaction to SDI—The Debate on the Nature of Deterrence," in Brauch, ed., *Star Wars and European Defense*.
18. Hugh De Santis, "SDI and the European Allies: Riding the Tiger," *Arms Control Today* 16, No. 2, March 1986.

determine the course of French thinking about what form an ATBM system should take. The emphasis now is on terminal defense of valuable military targets, with the heavy emphasis on kinetic or explosive kill systems capable of intercepting missiles in low atmosphere.[19] ATBMs have reverted to a purely military issue, and they are no longer being portrayed as a technological catalyst for European economies.

United States

The United States also has a severe budget problem which is increasingly placing constraints on military spending. The Congress—Democratic since 1986—has shown that it is unwilling to fund SDI to anything like the extent the Reagan administration wishes. In such a situation it is likely to be such Eurocentric systems as ATBMs which are reduced most severely in the American program. Although R&D seems set to continue on kill mechanisms and sensors suitable for the European theater, it is highly unlikely that Congress would be willing to spend the money to deploy an active defense system for the protection of European targets if Europe itself were not prepared to foot a large percentage of both the R&D and deployment bills. In addition, some of the enthusiasm for ATBM within the Reagan administration was based on a belief that Europe was the perfect place to test and deploy SDI-derived technology without clashing with the provisions of the ABM Treaty. As ABM Treaty restrictions only apply to technologies to counter strategic ballistic missiles, the development of technologies to counter theater ballistic missiles should remain unaffected. Because the problems of intercepting TBMs were at least as difficult as those involved with tracking and destroying ICBMs, the ATBM approach appeared to offer a perfect solution to many of the problems SDI testing was having with the ABM Treaty.

However, two things derailed this approach. Firstly, it became obvious that whatever its misgivings about SDI, Congress was prepared to allocate funds to the project. Hopes were not so sanguine about substantial ATBM funding. Secondly, the whole debate on interpretation of the ABM Treaty meant that there was at least the prospect of an alternative solution to

19. Alain Carton, "Extended Air Defense for Europe," unpublished manuscript, Institute for East-West Security Studies, 1987.

the SDI testing problem. Although willing to abide by the restrictive interpretation of the treaty for the time being, the U.S. administration has left no doubt that it believes the broad interpretation of the treaty to be correct. Furthermore, it gives every sign of being ready to move to a wider interpretation should its testing program require it. This knowledge was coupled with a strong urge on the part of the administration to proceed with early SDI deployment to protect targets in the United States itself. Such an early SDI deployment would utilize the technology which had been earmarked for ATBM. The need for ATBM, for the purely narrow SDI perspectives of the U.S. administration, was therefore gone—and with it much of the political pressure on European governments to press ahead with some form of concerted ATBM program.

4

Threat Assessments

NATO viewed the Soviet threat to its infrastructure in Western Europe as coming primarily from the new generation of Soviet nuclear-armed ballistic missiles. It was generally recognized that such missiles represent a qualitative improvement over previous Soviet capabilities in terms of accuracy. However, the estimated CEP of these missiles remains the subject of a considerable debate within Western defense intelligence circles, with estimates which range from 100 to nearly 900 meters, depending on the particular missile being discussed and the source being quoted.[20]

Soviet Missiles in Evolution

What is agreed is that these second- and third-generation missiles represent a major step in the evolution of Soviet missile design. The withdrawal of the SS-20, SS-22 and SS-23 which has been agreed as part of the INF Treaty reduces this threat but the SS-21, as well as the SCUDs and FROGs remain. Unless there are further limitations imposed as a result of future arms control agreements, current Soviet capabilities for highly accurate targeting will continue to improve as new generations of missiles are introduced.

The attempt, however, to present the military situation in Europe which emerged with the deployment of the latest generation of Soviet theater ballistic missiles systems as a dramatic shift from a previous position of relative NATO immunity to such threats is not one which can legitimately be

---

20. See Barton Wright, *World Weapon Database, Volume I: Soviet Missiles* (Lexington: Lexington Books, 1986).

asserted. The SS-4 and SS-5 were missile systems whose role was comparable to that of the SS-20. Even if their accuracy and reliability left much to be desired, sheer numbers would have ensured that their intended targets would have been destroyed, especially in light of the relatively high yield of their respective nuclear warheads. Thus, while the SS-20 did pose a qualitatively new threat, especially given its three MIRVed warheads, it was a situation which NATO had managed to live with over the past twenty years. As in most military spheres, the emerging threat from the Soviet Union was evolutionary rather than revolutionary.

Some West European countries, however, played up the threat posed by the SS-20 because they were concerned about the possibility of the political and military decoupling of the United States from the defense of Europe in the wake of the trauma of Vietnam and the phenomenon of Mansfieldism, or pressures for U.S. troop withdrawals, in the U.S. Senate. The SS-20 became a convenient hook on which to hang what were initially West German attempts, although later those of all the European allies, to strengthen the United States's military, as well as the political, defense links with Western Europe. The SS-20 threatened more targets more efficiently. However, it could still be argued that its use with nuclear warheads would, in the final analysis, still have been deterred by the American SLBMs allocated to SACEUR as well as by the SLBMs of both the United Kingdom and France.

The launching point for the current debate on ATBMs really came from the conviction that the improved accuracy of the SS-20 promised, in the foreseeable future, to give the Soviet Union the possibility of arming its new SRBMs with conventional warheads.[21] Such a configuration would have opened up the possibility of "surgical strikes" with conventional weapons on NATO military installations. For NATO, the fear was that such an approach would represent a more militarily attractive alternative for the Soviet Union than the option of initiating a

---

21. William M. Arkin and Jeffrey I. Sands cite a 1979 Tactical Air Command (TAC) threat briefing which states that "previously, Soviet SRBMs [short-range ballistic missiles] were considered primarily a nuclear threat. However, recent evidence indicates the Soviets may be developing a credible conventional SRBM capability against a variety of threats, including airfields." William M. Arkin and Jeffrey I. Sands, "The Soviet Nuclear Stockpile," *Arms Control Today* 14, No. 5 (June 1984), p. 5.

nuclear exchange of the type which the INF deployments of Pershing IIs and cruise missiles were designed first of all to deter and, if need be, react to. A crucial point for NATO was that because the decision was taken to concentrate on deterring a Soviet theater nuclear strike, the actual number of NATO ballistic missiles deployed in Europe did not attempt to match the numbers available to the Soviet Union. NATO judged that along with the 108 Pershing II missiles, 464 cruise missiles were adequate to deter a Soviet nuclear strike.

The conventional capability which some NATO planners had projected for the SS-20, -22 and -23 would have presented a variety of other problems for NATO. While all of these missiles were assumed to be armed with nuclear weapons, the numbers game in terms of pure arithmetic, as well as attempts to second-guess their targets, remained, if not simple, then at least a calculation well within the capabilities of NATO's military staff. However, if an undisclosed proportion of these missiles were to be armed with conventional weapons, and the Soviet Union also began to exercise more fully its existing option of deploying chemical warheads on some of its missiles, then the judgment made by NATO with regard to Soviet targeting intentions would become much more complicated. It may even have been impossible without data derived from hard intelligence on the various proportions of missiles allocated to each task. The INF Treaty does not resolve this dilemma. The treaty merely shifts the calculation onto the future configuration of the SS-21, SCUD and FROGs and their successor systems which will remain deployed in Europe.

NATO is well aware of its conventional inferiority with regard to the Soviet Union and its Warsaw Treaty allies. The full flowering of this debate has been recently revived with the Soviet proposals to eliminate all tactical nuclear weapons—the "triple-zero" solution—from Europe. The problem arises, and it remains so far a theoretical problem, of what happens if instead of dismantling nuclear missiles, a deal is struck to allow a proportion of them to be converted to carry conventional warheads. This is not a problem which is unique to NATO because such a proposition could equally affect the WTO if applied to the next generation of NATO tactical ballistic missile designed to replace the dual-capable Lance missile. If a sufficiently powerful conventional warhead were to be developed and combined with the ability NATO has already demonstrated with the warhead of the Pershing II to penetrate hardened targets, then these missiles would have the capability to threaten Warsaw Pact C^3I installations and air fields, as well as Soviet TBM launch sites, troop concentrations and supply dumps in Eastern Europe.

The question which must be addressed is what new qualitative threat improved technology has created. The significant improvement which took place in the CEP of Western long-range INF forces presented the Soviet Union with the same dilemma as the West faced from the introduction of the SS-20. By increasing the accuracy of these missiles, the destructive capability of a nuclear weapon is multiplied, while at the same time making it possible to lower the yield of the missile's nuclear warhead, thus lessening the risk of collateral damage due to the need to have large warheads capable of inflicting damage over a wide area. However, the reduction in the explosive yields of nuclear warheads also raised on both sides the specter of "surgical strikes" which could decapitate command and control networks as well as selectively remove weapons from the other side's arsenals. The INF Treaty removes this threat. However, the next logical progression is that the improved explosive yield of conventional munitions could make possible non-nuclear weapons with similar destructive capabilities, utilizing, perhaps, fuel-air explosives. When allied with the improved accuracy of the new generation of missiles, it could make them capable of carrying out what had previously been nuclear tasks.

■ *Soviet Capabilities*

For NATO's assessment of the level of threat posed by the possibility of Soviet conventionally armed TBMs, the most basic question which must be asked is what the Soviets may actually be able to be deliver, in terms of conventional explosives, utilizing existing or future generations of TBMs. Estimates vary widely, from over one metric ton of TNT equivalent per missile down to 200 kilograms. The lower end of the estimates seem more plausible, if allowance is made for the space taken up in a warhead for a missile's guidance package. Nor does it seem possible for the Soviets to substantially decrease the amount of space allocated to the guidance system, as it is this package which provides the level of accuracy which makes the deployment of conventional warheads a worthwhile proposition.

Whichever figure is taken for the theoretical conventional throw-weight for existing and putative Soviet missiles, all compare very unfavorably with the multi-ton capability of any of the tactical aircraft deployed by Soviet Frontal Aviation, not to mention the capabilities of Soviet Long-Range Aviation forces. In addition, whatever the attrition rate which NATO's air-defense capabilities may be able to visit upon Soviet Frontal Aviation, the vast majority of these aircraft will still fly more than one mission and deliver more than one payload before they are lost.

The suggestion has been advanced that if NATO achieves its projected 20- to 30-percent attrition rate for each Soviet sortie, then there may come a point where TBMs look like an attractive option for conventional use because there will be no Soviet pilots left, either due to loss or unwillingness to fly, to deliver aircraft-borne munitions. This belief, however, not only ignores the damage which would be done by aircraft before any serious degradation of their numbers took place, but it also underestimates the strength of Soviet military discipline and the willingness of soldiers, both individually and collectively, to put themselves in harm's way regardless of the objective chances of survival. Added to this, aircraft offer a level of flexibility not available from TBMs. TBMs are very much a one-shot option. After the reductions called for in the INF Treaty, around 1,500 Soviet TBMs will be available for both nuclear and conventional use. Even if this number increases to partially compensate for the loss of SS-20s, -22s and -23s caused by the INF Treaty it is unlikely to increase beyond the point where around 500 missiles, between a third and a quarter of the total, is available for a conventional strike. It does not seem likely that their delivery capability will exceed more than a hundred tons of conventional high explosives. Given the questions that remain over the accuracy of Soviet missiles, both now and in the future, it appears that conventionally armed TBMs still do not offer the prospect of the same level of punch or flexibility currently conferred by manned aircraft.

If only a relatively small tonnage of TBM-delivered conventional explosives are likely to be available to the Soviet Union, what is the most useful application for this limited capability? The most often mentioned use for such a capability has been as defense-suppression weapons. Such a use would place the focus for the use of such weapons upon attacks aimed against NATO's air-defense system, as part of a preemptive attack heralding an all-out WTO offensive into Western Europe. Soviet conventional TBMs would, in this scenario, be used to attack NATO's air-defense missile batteries and air-defense radars, in addition to being used against NATO's twenty-five main air bases. The primary objective of this type of attack would be to temporarily disrupt the operation of such air fields, with the additional objective of destroying as many aircraft on the ground as possible.

Such TBM attacks would be quickly followed by aircraft strikes on these same air fields, which, in an ideal world, would be unable to scramble their own aircraft for self-defense due to the closure of their runways. Soviet aircraft would, of course, have approached their targets through a NATO air-defense system which had been badly mauled by the same type of conventional TBM attack. By using TBMs to destroy air-defense radars, safe corridors would have been torn through NATO's air-defense missile belt in the FRG. Soviet aircraft would be free to operate in a less threatening environment and many more such aircraft, up to two-thirds of the first wave of Soviet Frontal Aviation, would be available for alternative targets and roles—in particular for ground support—because they would no longer be required for radar-suppression tasks.

However, a skeptical eye needs to be run over both the practicality and utility for this use of conventional TBMs. Calculations of the number of conventional TBMs which would be available to the Soviet Union therefore become critical in assessing the validity of the threat assessment.

It has been estimated that, given the ground support infrastructure necessary for TBM deployment in time of crisis, only around 300 TBMs would be available for an initial, preemptive attack. As a proportion of these missiles, probably around 25 percent, would have to be kept in reserve for a contingency which would require the use of nuclear warheads, the number available for conventional use drops to around 225 missiles. Given the long list of NATO targets available, embracing, *inter alia*, air fields, radar sites and air-defense missile batteries, it does not leave many missiles per target. In addition, there are problems with the type of scenario which forecasts the preemptive use of conventional TBMs. The successful use by the WTO of conventional TBMs in such a scenario depends heavily on the idea that the WTO would attack from a standing start, without NATO receiving any worthwhile warning. However, NATO planners today have largely discarded this type of worst-case scenario as a basis for serious planning forecasts.

■ *Timing and Damage*

The main reason why such a scenario is no longer at the forefront of operational planning stems from a widespread belief within NATO's military leadership that, whatever the initial success of the Group of Soviet Forces in Germany (GSFG), its formations would quickly burn out without rapid reinforcement. Such reinforcements would require at least partial prior mobilization by the Soviet Union, which NATO is confident of observing. Currently, therefore, all of NATO's military and political planning for a conflict in Europe presupposes a certain amount of warning time—this warning time being intended to enable NATO's military forces time to mobilize and take up war stations. In terms of the conventional TBM threat, it should enable NATO to implement the extensive dispersal plans it has for its air assets. Such dispersals would greatly complicate the targeting problems faced by the Soviet Union in deciding whether it should launch its conventional TBMs. The Soviets would be faced by many more targets, with each individual target being of less value than each of the original target set. If the Soviets stay with their original target set, and always assuming that NATO's dispersal plans work, then a significant proportion of NATO's air-defense assets would survive the attack untouched. Thus the first wave of Soviet aircraft would face the prospect of being detected by reserve radar stations, attacked by air-defense missiles which have redeployed and met by NATO fighters rising from secondary air fields.

Added to these issues, there is also the important question of timing which surrounds the possible use of conventional TBMs in the defense-suppression role. The flight time of the typical Soviet TBM is at present around five minutes. If Soviet aircraft are to take full advantage of the confusion wrought by a TBM attack, they need to arrive over their own targets as soon as possible after the missiles have hit, while the dislocation of the NATO air-defense environment is at its maximum. This would mean crossing the Hawk belt of air-defense missiles not more than five minutes after the TBM strike and arriving over the air field, the primary targets, five to ten minutes later. However, the flight time to these targets for many of the Soviet fighter bombers which would be used in such tasks is around forty-five minutes. This would mean that such aircraft would have to begin taking off forty minutes before the TBMs were launched.

NATO's air-defense radars, in particular its AWACs, would be bound to notice such a massing of Soviet aircraft and, in all probability, would also be able to divine Soviet intentions. NATO can scramble many of its aircraft in thirty minutes, and it may therefore be the case that the Soviet Union would be willing to forego the benefits of its aircraft arriving quickly after the TBM attack in order to keep NATO's aircraft on the ground. This would also allow a proper damage assessment to be carried out by the Soviet Union and give its air crews some idea of what sort of an environment they would be likely to meet over Western Europe. However, forty or fifty minutes gives NATO time to draw its breath. Warnings by AWACs that Soviet aircraft were massing, no matter how threatening such a situation appears, would not galvanize NATO in the way that a TBM attack would. If conventional TBMs only wound NATO, without doing it mortal damage, then such an attack may do the Soviet Union more harm than good. It would truly have wakened a sleeping tiger.

It therefore comes down to a fine calculation of the damage likely to be done by conventional TBMs. The figures, especially from the Soviet point of view, are not very impressive. The weight of conventional munitions which it seems possible that TBMs could deliver are minuscule compared to the capabilities of Soviet aircraft. With figures of 300-meter CEPs for such missiles as the SS-21, it seems unlikely that the high degree of accuracy needed for a preemptive conventional TBM strategy can be consistently relied upon. If, as may be possible in the case of the SS-21, accuracy can be improved by adding terminal guidance to existing initial guidance packages, then there is a price to be paid in terms of reducing the missile's explosive payload. This means having more missiles available to inflict much the same level of damage. Economic costs to one side, the arms control perceptions involved in substantial new missile deployments may place such an approach beyond any current Soviet agenda.

NATO is not alone in making such assessments. The Soviet Union has also expressed its concerns over the increased capabilities of NATO missiles, especially as they perceive Western technology in guidance systems and the chemical sciences, in the shape of new explosives, to be constantly outstripping their own efforts. There is, therefore, an element which runs through the military thinking of both alliance systems which sees the very real possibility of conventional ballistic missiles being used preemptively for a first strike either in support of or directed against a conventional attack in Western Europe.

From the WTO point of view there is the risk that the temptation for NATO to launch a preemptive strike will no longer be inhibited by the political odium which would inevitably surround the first use of nuclear weapons. For NATO, despite political declarations to the contrary, it could reinforce an existing belief in the attachment to the concept of preemptive strikes on the part of the Warsaw Pact. It would also, in NATO eyes, represent a significant enhancement of the capabilities of the WTO countries for launching such an attack. NATO could well regard the deployment of conventionally armed ballistic missiles by the WTO as being designed to reinforce the belief that NATO would find it very difficult to respond with nuclear weapons to even the most disarming conventional strike. Such a step, the WTO might calculate, is unlikely to be sanctioned until it became obvious to both political and military leaders that the actual conventional battle in central Europe had been joined and was demonstrably being lost by the West. It is, therefore, necessary to keep these perceptions in mind when exploring the usefulness, both perceived and actual, of active defense as a means of countering this problem, both as an approach into its own right, and as part of a mix of active, passive and collateral measures.

5

The Impact of the INF Treaty

The INF Treaty has had an obvious impact on the ATBM question. The treaty removes a whole class of weapons which could have, if converted to conventional use, provided a major rationale for the deployment of active defenses. However, after the initial withdrawal of existing nuclear systems, what are the longer-term implications of the treaty? For the Soviet Union, the destruction of the Pershing II missiles deployed in the Federal Republic removes a real possibility of conventional attack on infrastructure targets. The Pershing IIs' forty-meter CEP made it the only serious candidate for a conventional role among the existing generation of both American and Soviet theater missiles. For NATO the withdrawal of SS-20, -22 and -23 will have little immediate impact on its ATBM discussions. All of these missiles have come to be discounted as serious contenders for a conventional role. Once the exaggerated initial fears over their improved accuracy had been allayed, they were viewed as primarily nuclear systems, with, perhaps, in the case of the SS-23, some residual conventional role. Attention in NATO was focused rather on the successor systems and the possibilities which they might offer the Soviet Union for a realistic conventional capability.

It is the disappearance of these missiles from any future Soviet inventory which is the major long-term implication of the INF Treaty for the ATBM issue in NATO. An entire class of missiles no longer needs to be countered, as the INF Treaty covers not only future nuclear but also conventionally armed intermediate-range missiles. The other major impact is on the residual force structures which will ensue from implications of the INF Treaty cuts in the numbers of missiles deployed. Clearly, many of the targets allocated to INF missiles will now be placed on the shoulders of the remaining SRINF missiles. This is likely to be the case in both alliances. However, this situation places both alliances in a quandary which has a direct impact on the utility of any future ATBM system. At present NATO has available eighty-three Lance missile launchers, fifty-seven belonging to European countries and thirty-six to the United States.[22] These launchers and their two, three or four missiles apiece constitute what is now available as a substitute for American Pershing II, cruise missiles and the West German Pershing IA, which are also being taken out of service as a consequence of, although not as a provision of, the INF Treaty.

These Lance missiles remain NATO's visible missile-borne deterrent in Europe. Aircraft and sea-launched missiles are not felt, at least by politicians, to exercise the same deterrent effect as land-based missiles. (This was a major rationale for introducing Pershing II and cruise missiles into Europe). As a result of this perception, and West German misgivings notwithstanding, it is extremely unlikely that any of these Lance missiles, despite their dual capability, are ever going to be available for conventional use. There are simple too few of them and too many potential targets for a nuclear strike, even within their restricted range of 110 kilometers.

The moves to provide a modernized successor system to Lance, with greater accuracy and longer range, has sparked an argument in NATO regarding the next step to be taken in arms control. The West Germans, worried that only Germans, East and West, will be the targets for the remaining short-range nuclear missiles, wish to begin immediate negotiations for their elimination—the "triple-zero" solution. Both the Americans and the British, however, have insisted that short-range nuclear systems can be dealt with only after reductions have taken place in conventional forces in Europe and chemical weapons have been eliminated. British Prime Minister Margaret Thatcher has categorically stated: "[There is] no case for further nuclear weapons reductions in Europe before a balance

22. These figures are derived from the International Institute for Strategic Studies, *The Military Balance* (London: IISS, 1987).

in conventional forces and a ban on chemical weapons has been negotiated."[23]

The United States and the U.K. may be able to prevent any reduction in the number of nuclear missiles deployed in Europe. They may even be able to force through the full modernization program outline in 1983 and deploy more accurate missiles with ranges up to the 500 kilometers allowed by the INF Treaty. However, the strong opposition in West Germany, plus political and public sentiment in the rest of Europe, will make agreement on an increase in the number of missiles deployed unlikely if not impossible.

This refusal to deploy any additional short-range missiles may be the crucial fallout of the INF Treaty. A large segment of public opinion in Europe seeks continued reductions in the number of nuclear weapons deployed on the continent. Nuclear weapons have come to be almost wholly identified in the public consciousness with land-based missiles. Despite nuclear-capable aircraft deployed throughout Europe and nuclear-armed submarines patrolling its contiguous seas, it is land-based missiles which have become the rallying point for both Western publics and politicians. Prime Minister Thatcher might not rule out reenforcing the number of U.S. F-111 bombers in the U.K. or basing cruise-missile-armed submarines in Scotland. But the U.K. is unlikely to deploy more missiles, and the same is true for the West Germans. They may modernize existing missiles, and, as suggested by Norwegian Defense Minister Johan Jorgen Holst, describe it as "force improvement."[24] However, the Germans are likely to balk at the idea that numbers need to be increased. In sum, there is no realistic prospect of deploying more short-range missiles than the number of Lance missiles currently deployed. Therefore, NATO will have no spare capacity to use conventional ballistic missiles even should it deploy systems which are capable of performing this role.

What does all of this mean for the ATBM debate? The WTO does not face a shortage of short-range nuclear systems. It has been claimed that they have 1,400 such missiles available on the central front. But looked at more closely, the Soviet Union and its allies face exactly the same problems as NATO. The

---

23. *Financial Times*, London, February 18, 1988, p. 2.
24. *The New York Times*, February 3, 1988.

WTO's impressive total of short-range nuclear missiles is made up largely of aging SCUDs and FROG missiles, which, like Lance, are approaching obsolescence. Even the most modern remaining missile, the SS-21, is only two years younger than Lance. Furthermore, the Soviet Union has only 130 launchers, with four more each deployed by the GDR and Czechoslovakia.[25] It is the replacement for this missile which is the most obvious candidate for the conventional role. However, the Soviet Union places a heavy reliance on its short-range nuclear forces—a reliance which can only increase now that its INF force is set to disappear. It is hardly credible that the Soviets would be willing to see a majority of their new missiles given over to the conventional role. Nor, in a continent which has grown missile-shy, is it practical to consider massive new missile deployments to provide the numbers necessary to perform both the conventional and the nuclear role. Conventional warheads may simply be left to rust because no one is willing to spare a launcher or missile for the task. Without a credible large-scale conventional TBM threat, what remains of the rationale for pursuing ATBMs?

---

25. *The Military Balance*, 1987.

6

Active Defense and the ABM Treaty

The entire debate surrounding ATBMs for Europe is carried out in the shadow of the ABM Treaty. Much of the political debate in Europe on ATBM hinges on what is permissible under the terms of the ABM Treaty. Central to this debate are the types of technology transfers of ATBM- and BMD-relevant systems, both in terms of hardware and software, from the United States to Europe which may be acceptable under the ABM Treaty. Western Europe as a whole has pronounced itself as being broadly supportive of the restrictive interpretation of the treaty's provisions.[26] Retaining this approach would, in all probability, place considerable restrictions on the type of SDI-derived technology which might be made available for any European ATBM system. No current or foreseeable government in any major European country would be comfortable being involved in technology transfers from the United States which vocal sections of both its own domestic public opinion, and an important section of U.S. political opinion—as well as the Soviet Union—were claiming were in breach of the provisions of the ABM Treaty.

■ *What is Covered*

In this regard, the whole question of definition of what systems are covered by the ABM Treaty and which are excluded becomes very important. It must also be borne in

26. Michael Lucas, "SDI and Europe," *World Policy Journal* 3, No. 2 (Spring 1986).

mind that tactical ballistic missiles are not covered by the provisions of the ABM Treaty. Their exclusion is based on the fact that they are not strategic in range under the criteria for ranges agreed between the Soviet Union and the United States. However, it is also important not to overlook such SLBMs as the Soviet SS-N-5. These missiles are judged to be strategic, although their range, at only 1,400 kilometers, is less than that of a theater-range weapon such as the SS-20. This is because for the purpose of SALT and the ABM Treaty, submarine-launched ballistic missiles were all treated as being strategic in range. Because the SS-N-5 has much the same trajectory as the SS-22 the handling of a technology which has this type of double characteristic is obviously a critical grey area when it comes to deciding whether the ABM Treaty could apply to an ATBM system. It is this type of question which is likely to be most critical in terms of the type of technology which could be transferred from the United States to the Europeans and the type of technology which the Europeans might feel more comfortable developing for themselves, should NATO decide to pursue active defense.

There are thus two conflicting interests when it comes to the subject of ATBM which complicate the picture when deciding which type of defense system can best serve West European needs. While there may be strong arguments in favor of a terminal-defense system based on some form of air-defense missile upgrade, a major consideration in any attempt to assess the likely effectiveness of such a system would be the type of sensor systems which would have to accompany this form of terminal defense. A full range of space-based sensors would probably be necessary to make a ground-based missile, designed for endoatmospheric interception, a realistic possibility. To depend on ground-based radars to acquire the target and track it for a sufficient amount of time to provide data for interception is possible and a realistic option, even given the relatively short flight times of the ballistic missiles involved.

If it is decided to go the route of air-defense upgrade to counter the TBM threat, this may even represent the most probable initial configuration of any ATBM radar system to be deployed. However, sensor systems able to acquire missiles at launch and track them from the beginning of their trajectory would be highly desirable. This capability would allow additional time for calculations to be done which would enable missiles to be on their way to intercept the ballistic missile warheads as they reenter the atmosphere from space. It is in this area that the clashes with the ABM Treaty are likely to occur.

■ *West European Views*

It is difficult to conceive of Western Europe being able to afford to develop this type of sophisticated sensor technology on its own, even in the absence of the SDI program. The fact that these types of sensors are already being researched as part of SDI takes away most of the political incentives for spending scarce resources on this type of research. West European governments could plausibly argue that this type of duplication in R&D is both unnecessary and undesirable. By pursuing independent research in this field European countries would feel themselves to be falling into the trap which they have sometimes accused the United States of flirting with—namely, devoting too many scarce resources, both in terms of manpower and monies, to esoteric research.

West Europeans consider that such research is more often than not totally unconnected to the more immediate and pressing need to upgrade NATO's conventional capability. Although such a capability is primarily intended for the defense of Europe, the Europeans would argue that it also offers benefits for the United States, especially the American ability to project military power into other regions of the world. The West Europeans are likely to argue that it would be much better for NATO if they were to devote more of their resources to the myriad types of emerging technologies (ET) and other conventional weapons systems which Western Europe is likely to require in what promises to be an increasingly denuclearized environment. It is here that the heart of the European dilemma lies: an unwillingness to forego an attractive technological option, paid for by someone else, pitted against political and public opinion constraints on using SDI-derived technology. The temptation may become too strong to resist. If an ATBM system is agreed to be desirable then the West Europeans may find themselves, because of budgetary pressure, queuing up to utilize American technology for a West European ATBM system.

This approach, however, involves numerous difficulties. It presupposes a willingness on the part of the U.S. administration and Congress to make the fruits of their research available to the West Europeans at an affordable price. Given the perennial problems over burden-sharing within NATO, there is likely to be strong opposition to this from Congress. But essentially it would most probably come down to a matter of political haggling over the price which Western Europe will have to pay to have access to the results of American research. While this issue is mainly an intra-alliance issue, the most serious problem is the implications of the transfer of this type of U.S. technology to Western Europe in terms of the permissible technology transfer under the ABM Treaty. There is no doubt that the Soviet Union would seek to exploit possible ambiguities in the treaty and would attempt to embrace all possible sensor technology, as well as kill mechanisms, within the definitions agreed under the treaty. The West Europeans would find themselves on the horns of a dilemma should this become, as seems inevitable, a sensitive political issue.

All West European governments have repeatedly emphasized that they wish to see the ABM Treaty maintained and that they believe it remains the cornerstone of the only remaining strategic arms control regime, one to which they attach great importance. They also view it as the bedrock for preventing an arms race in outer space. Furthermore, West European governments have also indicated their support for the original, restrictive interpretation placed on the treaty by the Soviet Union and by former U.S. administration officials. There has been little sympathy shown by West European governments for the broad interpretation of the treaty's provisions.

It is likely that there would have to be significant and tangible payoffs in terms of both enhanced public perceptions of increased security and improved inter-alliance relations in order for them to be prepared to pay the undoubted price, both in terms of their relations with the Soviet Union and also with considerable, and vocal, sections of their own public opinion, in accepting the broad interpretation of the treaty and any transfer of technology which could plausibly be claimed to have been prohibited by the restrictive interpretation. Once again political considerations conflict with fiscal and military realities. Solutions will depend on two threat assessments. One is the domestic turmoil for all the parties involved which may result from transfers; the other, the threat assessment of Soviet capabilities and intentions which provide the motor for weapon procurement decisions. A less offensive Soviet posture and strategy could at a stroke remove much if not all of the incentives for proceeding with a difficult procurement decision.

It is into this highly charged environment that the Americans have attempted to inject their own concern about the stability of the nuclear balance at both the strategic and the European theater levels. The whole debate surrounding SDI and its impact on strategic stability is outside the scope of this work, but SDI has acted as a major catalyst for a reexamination of the offense/defense relationship and it is the place of ATBMs in this evolving relationship which need concern us here. An examination of the most likely deployment scenario for active defenses and its possible impact on strategic stability in Europe would therefore repay close examination.

■ *A Possible Scenario*

This scenario is based on the assumption that either NATO or the Warsaw Pact could, for a time, have a monopoly on ATBMs after a unilateral deployment. Such a situation would obviously have a major impact on perceptions of stability in terms of the other side being able to initiate either a successful first strike or even a retaliatory second strike. Although the effectiveness of the ATBM system may be open to question, its very existence would complicate the Euro-strategic equation. For example, C^3I installations could be at serious risk from conventionally armed ballistic missiles. Deployment of ATBM around such installations by one side, while the other side's installations remain at risk, seriously decreases the ability of the undefended side to anticipate waging war effectively.

However, on closer examination of this scenario, the idea that there could be a unilateral deployment by, for example, the West is not one which holds up to serious examination. The Soviet Union has an extensive BMD research program which is the equivalent in many spheres of that being carried out by the United States. As actual deployment of either strategic or theater BMD is likely to be partial and incremental, there are strong incentives on the part of the other side, whoever the other side happens to be, not to allow any unilateral deployment to go unchallenged.

To take a possible example, if the Soviet Union were to go ahead and deploy an ATBM around its C^3I installations in the western Soviet Union, there could be strong political incentives on the part of NATO to also rush into a token deployment of its own ATBM in order to satisfy the idea that there should be equality in the capabilities of both sides and that neither side should be allowed to gain a unilateral advantage. The fact that the second deployment may in fact be introducing a system which is not technologically perfect or operationally efficient is less important than the political symbolism. The reason for this is that the actual deployment of an ATBM system is likely to be, itself, an act of political symbolism or, at best, deterrence by denial. It would most probably involve a technology that is far from mature and which may be a considerable way from having even marginal effectiveness.

It is therefore likely to be the case that whoever deploys first is going to do so by only a hair's breadth, and he will not derive much, if any, military advantage from the deployment due to the inadequacies of his own system and the prospect of new simultaneous deployment by the other side. Given the worst case, and this is the approach which all military planners seem to indulge in, there will no doubt be, once again, an ever-spiraling assessment of the capabilities of the opposition, coupled with a prolonged debate over the shortcomings of one's own system. The period of advantage for the side who deploys first could be all too ephemeral. This being the case, the focus then shifts onto ways of negating or defeating the ATBM system of the opposition.

7

Countermeasures to ATBM

The obvious response to the deployment of an ATBM system is to increase the number of offensive missiles in order to overcome any possible defense by simply swamping the system in terms of either its target-acquisition and tracking capacity or overcoming by sheer weight of numbers the kill mechanisms available to the defensive force. However, this is not the simple option it first appears. Unless detailed and accurate information is available on the capabilities of the defensive system, it is hard to judge what level of offensive forces would represent a number sufficient to ensure that the defensive system was effectively swamped. In addition, the deployment of large numbers of offensive systems would be perceived as highly destabilizing and provocative. Such an approach would also be viewed as totally contrary to the professed aims of both sides in terms of their desires for arms control and the deep reductions in their respective offensive ballistic missile forces.

For these reasons a number of other alternatives are being explored in terms of preferred countermeasures to ATBMs. These alternatives are being examined to identify what can be done to protect offensive ballistic missile forces and to assess what sort of opposition, in terms of future BMD systems, may have to be overcome. These options embrace such approaches as shorter boost phases for missiles, increased numbers of decoys, better cloaking of real warheads among decoys, shorter flight times, as well as improving the MARVing capabilities of individual missile systems. These and other alternative countermeasures may be relatively cheap when compared to the expensive undertaking of building and deploying additional ballistic missiles. They could represent a cost-effective alternative to a straight numerical buildup in offensive forces both in terms of their economic and political impact.

However, it must be realized that the uncertainties which surround this approach to countermeasures, in terms of political, if not technical, perceptions are likely to be greater than those around a straightforward numerical increase in ballistic missiles. Politicians and public opinion are notoriously unwilling to grapple with the finer technical points of weapons technology. There is a strong, almost atavistic, tendency on both sides to go for raw numbers because a general perception exists that security comes with higher numbers. The fact that so much of this technology is cloaked in secrecy and is technically at the leading edge of science makes it difficult for both the public and politicians to grasp what the debate is actually about. Although military and scientific experts may make a telling case for going the road of particular passive countermeasures to an ATBM system, it is questionable, given the present level of debate, how much impact such an approach would have on political decision-making. This may be particularly true in the Soviet Union, which in the past has displayed a marked tendency to go for buildups in terms of raw numbers in order to compensate for the perceived technical proficiency of the West. It may be difficult to convince the Soviet political and military leadership that their offensive countermeasures technology is equal to the task of overcoming the West's new defensive technology.

It is certainly true that passive countermeasures, no matter how effective, have little payoff in terms of greater political leverage in influencing the debate on the arms race now going on in the West. Such an approach would give little ammunition to those groups who oppose both SDI and ATBMs because they view them as fueling yet another spiral in the arms race. The opposite may be the case for the United States, where there seems to be greater confidence in the ability of American scientists to overcome, through technical fixes, any Soviet deployment of a BMD system. This approach seems largely to originate from a general tendency in some important administration circles to deride the capabilities of Soviet military technology generally. It is questionable, however, if such xenophobic opinion can be maintained today given both the indigenous achievements of Soviet science and also the Soviet

Union's well-known capability for obtaining critical defense technology from Western sources.

However, there is no doubt that the situation in which both sides were rushing to deploy countermeasures would represent a major instability in terms of the overall strategic relationship between the two alliance systems as well as posing a serious strain on the European theater military situation. The availability of conventionally armed ballistic missiles which could be used for a first strike against significant military targets on either side is in itself destabilizing. A situation where such systems could be countered by a preemptive capability based on either the same type of ballistic missile technology or some improved form of stand-off air attack, coupled with the enhanced terminal defense of critical targets, all makes for a situation where the tendency to go for early and massive use of such conventionally armed ballistic missiles may be highly tempting. The rationale for a first strike in this situation would be to ensure the destruction of vital targets, protection of one's own ballistic missile assets and also a strong desire to test the efficiency of the other side's defensive technologies in order to form early conclusions about the usefulness of this technology, both offensive and defensive, in other elements of the conventional and nuclear battlefields.

8

Active Defense Choices

The ATBM debate has always been confined to a relatively small constituency. It never caught the public imagination in Europe the way SDI did in the United States. Discussion in Europe began with the same grandiose ideas which originally fueled the SDI research program. Talks of layered defense with space-based elements for mid-course or even boost-phase interception peppered the literature, especially that portion which emanated from the United States.[27] Those elements in the American government and research community which continue to view the deployment of ATBM in Europe as a testing ground and technology leader for SDI continue to think in terms of directed-energy-weapon (DEW) kill mechanisms and sophisticated target-acquisition and kill-assessment sensors. In terms of sensor systems necessary to effect interception of tactical ballistic missiles there is probably much in what they are thinking: it is certain that SDI-related, and probably SDI-derived, technologies would contribute greatly towards improving the effectiveness of any European TBM system. However, the low-key debate which is now taking place in Europe on the type of ATBM which the Europeans might choose to deploy is rather more prosaic. This discussion focuses instead on either upgrading existing air-defense missiles or building new types of missile systems based on existing air-defense technology but with enhanced speed and accuracy. It also places a much greater emphasis on ground-based rather than space-based sensor technology to

27. Gregory Canavan, "Theater Applications of Strategic Defense Concepts" (Los Alamos: Los Alamos National Laboratory, 1985), LA-UR-85-2117, June 4, 1985.

effect interception of TBM warheads at endoatmospheric heights rather than exoatmospheric range.

When looking at the actual technology involved in the interception of the variety of Soviet missiles which are judged most likely to pose a threat to important European targets, it is necessary to bear in mind not only the question of technical, but also political, feasibility. A variety of options exist or are in prospect for the interception of ballistic missiles which run the whole gamut of approaches which are currently under exploration in the SDI research program. As with the debate surrounding strategic BMD, initially one of the major questions in Europe centered around whether boost-phase or mid-flight interception of such missiles was technically feasible.

It was quickly realized that all the problems inherent in a full BMD defense system applied equally to an ATBM system. By attempting to attack the missiles during the boost phase or in mid-flight, the research was seeking to address the most difficult technical problems at a very early stage. It may therefore be the case that, for NATO, the real choice lies between defense prior to launch, by means of a preemptive strike on the launch before the opposition's missiles get underway, or attacking the missiles during their terminal flight phase, with active defense of the probable targets. The decision may therefore come down to a relatively simple choice between destroying enemy missiles on the ground, while they were still on their launch platforms, or in their final moments of flight.

Initially, as a result of the Counterair 90 study and the two AGARD studies, AAS 20 and AAS 25, carried out in the early 1980s, the emphasis of NATO's internal discussion on how to deal with the putative Soviet TBM threat focused on the destruction of Soviet missiles before launch. The technology needed for this is already available in the form of accurate ballistic missiles, such as Pershing II, which have the potential to be armed with powerful conventional warheads.

While technically feasible, the political constraints upon such an approach are considerable and in many ways they negate the technical advantages which NATO has. While the Soviet Union enjoys an internalized political decision-making process which would require only the agreement of the Politburo and perhaps the Soviet military to secure the launch of Soviet ballistic missiles, the position of NATO is considerably different. Before such a launch of missiles under NATO command could take place it seems likely that the whole process of securing permission from the sixteen constituent NATO members would have to be gone through. Whatever the depths of the political crisis, it is highly unlikely that a majority of NATO countries would be prepared to give carte-blanche approval to NATO's military command. In fact, the deeper and more serious the political crisis, the less likely are NATO political leaders to abrogate their responsibilities and place the burden of initiating a preemptive strike on Soviet targets, a strike which would inevitably mean full-scale war, on the shoulders of a military commander. Such a step represents the final political escalation of a crisis and it is a decision which political leaders will jealously retain for themselves. It is difficult to envisage, especially in the absence of overt Soviet military aggression, the politically diverse governments which form NATO being prepared to take the drastic step of being seen as the aggressor in a conflict.

Taking all this into consideration, the options for NATO then become boost-phase, mid-flight or terminal interception. The first two options involve the use of the very highest technology, both in terms of kill mechanisms and target-acquisition systems. All this technology is closely allied to the American SDI research program and its research, development and deployment share common difficulties with those involved in any ballistic missile defense against strategic ballistic missiles. The early results of the SDI research program suggest that a workable defense system against ballistic missiles which involves interception taking place during boost- or mid-phase appears to be decades away. Thus, such an approach to the problems posed by conventional ballistic missiles for the defense of targets in Europe is not a realistic short- or mid-term proposition.

■ *Terminal Defense*

The elimination of preemptive, boost-phase and mid-course interception as serious or immediate approaches to the ATBM question means that, at least for NATO, the main remaining option revolves around terminal defenses. Such an approach would have the advantage of relative technical simplicity, allied to a lack of political problems. It would furthermore be likely to involve, at least initially, the defense of only the most vital military targets located in Western Europe. The kill mechanism for such a system could involve an improved air-defense system, such as that promised by some form of "Super Patriot" missile. Such a kill mechanism is easily within the reach of existing U.S. and West European technology. Furthermore, the target-acquisition, tracking and kill confirmation for terminal defense is either already available, in terms of existing air-defense radars and AWACs, or could be affordably developed.

In addition, it might be possible for such a system to reap the early benefits of SDI research, initially in terms of sensor systems but later also in terms of kill mechanisms. SDI research at present is increasingly focused on kinetic kill mechanisms and also high-energy, ground-based laser systems. Both of these approaches to the destruction of ballistic missiles lend themselves easily to future integration into an existing ATBM system based on endoatmospheric interception and destruction down to low-atmospheric levels. Also, another strand of weapons development, epitomized by the emerging technologies available for conventional defense, could also be harmonized into such an approach to the problems of the interception of theater ballistic missiles. Once again, the utilization of this technology would create fewer political problems for NATO in any possible program to deploy a European ATBM system. How then might such an ATBM system be created?

■ *Patriot: A Case Study*

A whole variety of missiles have been discussed for use by NATO in the terminal-defense ATBM role. In addition to the Patriot air-defense missiles, the French Aster and Israeli Rafael missiles are also being touted as serious prospects if NATO decides that it is worthwhile to aim for a near-term deployment of ATBM. With this in mind, it is worthwhile to look at what advantages NATO might expect to gain from the deployment of this type of upgraded air-defense missile in the ATBM role and what the costs and disadvantages of adopting this approach might be.

The saga of the development and deployment of the Patriot missile system is both long and difficult. Originally intended as a replacement for the Hawk missile first deployed in the 1960s, Patriot had a very long gestation period, during which its costs escalated rapidly. By the time NATO moved towards deployment in the mid-1980s, it was estimated that it was costing around $250,000 per missile for Patriot deployment. Although Patriot is a far more capable air-defense missile than Hawk, NATO decided to follow a similar practice for the Patriot's physical deployment. Patriot missiles would be slotted into the existing Hawk air-defense belt which runs across Western Europe as the major NATO defense systems against Soviet tactical aircraft. Although deployment of Patriot in the point-defense role for high-value military targets was considered, the high costs of the missiles made this approach all but impossible by the time deployment actually began. The situation at present is that Patriot is being introduced piecemeal in the Hawk defense belt and those Hawk missiles which are being withdrawn are being considered for possible redeployment as point air-defense for such targets as air fields.

If Patriot is considered for use in the ATBM role, then there are two possible approaches which NATO can take to achieve this objective. The first is a relatively simple upgrading of the current mark of Patriot with the addition of new sensors, a new guidance system, new fusing for its warhead and replacement software for its command and control systems. The major difficulty with this type of upgrade lies in improving the radar tracking capabilities of the Patriot support radars to deal with the TBM threat. The reason for this is that a TBM presents a more difficult target than an aircraft because of its smaller radar cross-section and the fact that it comes in vertically rather than horizontally. However, this type of radar upgrade is still not too difficult a task to be fairly straightforward to solve, if sufficient funds are made available.

An additional incentive for this approach is that Patriot is itself a prime target for TBMs if ballistic missiles are to be used as a defense-suppression weapon before an air attack. This reason itself has been advanced as one of the advantages in pursuing this form of Patriot upgrade. It has been estimated that for less than $20,000 per missile it could be possible to give the Patriot system this form of self-defense capability. Thus, NATO's air-defense capability is protected for a relatively small investment.

The important point here, however, is that it would only be a self-defense capability. The "footprint," the area on the ground which an individual Patriot missile fire unit could reliable protect against TBMs, would be relatively small and the deployment of the Patriot missile in a belt means that only a relatively small area on the ground could be protected from TBMs. The present deployment of Patriot means that high-value NATO targets such as air fields and C^3I installations would all lie outside the field of coverage of any existing Patriot-upgrade ATBM capability.

For the oft-touted role of point defense of military targets utilizing only a marginally improved Patriot missile, what would be required would be an entirely new Patriot deployment, similar in scale to that currently taking place as part of the Hawk belt replacement program. These deployments would have to see Patriot missile batteries grouped around targets such as air fields or C^3I installations in relatively tight proximity, to enable their footprint to cover the bases themselves. Such deployments, however, would have to see Patriot batteries located in much closer proximity to air fields or other installations than was ever considered for point air defense. This older approach still placed the Patriot launchers some distance from the targets to be defended, to allow for additional time for target engagement, and would have left those targets outside Patriot's restricted ATBM footprint. But by doing this, it is argued, the Patriot missiles could perform a dual air-defense and ATBM role. While it is certainly the case that it might increase the protection afforded to air fields, the question then arises as to the cost-effectiveness of using modified ATBM Patriots for air defense. The use of such modified missiles against aircraft, as long as the TBM threat continued to exist, might be actively discouraged. This would tie up valuable air-defense weapons waiting for a TBM attack which might never emerge, or whose effect was marginal compared to the damage being inflicted by unmolested aircraft.

The other approach to the Patriot upgrade is to design an entirely new missile system and include an improved warhead and new guidance and booster systems along with an entirely new radar and software support package. This approach has very little to do with the Patriot that is now in service. It has, however, been promoted actively by those industrial interests concerned with producing the missile, along with part of the SDIO who seem keen to see the U.S. Army bear the cost of research and development for what is essentially a Eurocentric ATBM.

It is here that the bureaucratic politics of the American defense establishment play an important role. There exists a competitor air-defense/ATBM missile, referred to as the flexible lightweight agile experiment (FLAGE). A section of the U.S. Army bureaucracy is keen to see the FLAGE developed instead of Patriot because the cost of introducing it into service would be met from SDIO funds, and not from the U.S. Army's appropriation. The reason for this is that while Patriot is primarily an air-defense weapon, and hence an army responsibility, FLAGE is being presented largely as a potential BMD weapon, and hence it falls within the purview of the SDIO-funded programs. It may be on such considerations that the future shape of European ATBMs actually depends, rather than the grandiose strategic pictures which defense analysts are keen to paint.[28]

It is possible that an initial version of an improved Patriot could be available by 1993 but at some considerable cost to NATO. It has been estimated that each fire unit might cost in the region of $160 million, and it would be necessary to deploy two fire units for each of NATO's twenty-five air fields. In addition, while Patriot is a moveable missile in the sense that it can be dismantled and shifted to another site, it is in no sense mobile and there would be no realistic possibility for any flexibility in deployment patterns during a crisis. This lack of mobility also means that Patriot would not be able to provide any ATBM capability to major units maneuvering in the battlefield. Point defense by Patriot may offer some real benefits to NATO installations both in terms of the Soviet TBM and the Soviet air threat to these selfsame locations. However, it would be expensive to go through another round of Patriot deployment, even if adding on an ATBM capability is, in itself, relatively inexpensive. It might therefore be worthwhile to assess what other systems are available to NATO to meet the putative Soviet threat.

---

28. Hans-Guenter Brauch, "From SDI to EDI—Elements of a European Defense Architecture," in Brauch, ed., *Star Wars and European Defense.*

9

Passive Measures

The deployment of ATBM represents one reaction to the possibility of a threat emerging from conventionally armed ballistic missiles. However, it would be wrong to assume that development of active defense represents the only possible reaction to this perceived threat. A range of other alternatives also exists to counter or degrade the threat which conventional ballistic missiles could pose to the whole range of important military targets such as air fields, nuclear weapon storage sites, command headquarters, storage dumps and ECM posts, which are widely considered most likely to be prime installations under threat.

The most obvious response is the passive defensive measures designed to allow these military targets to ride out any attack and remain effective and fully operational after they suffered hits from conventional missiles. As the most obvious target set includes C^3I installations concerned with the command and control of military forces, the best way to effect passive defense of these installations is to ensure further hardening of the physical locations in which they are set, as well as increasing the amount of dispersal for important headquarters and intelligence-gathering and -processing units. Such dispersal would complicate greatly the target set for any possible ballistic missile attack. Improved mobility would provide further protection in terms of the possibility of quickly making short moves in order to render the targeting for a ballistic missile attack much more difficult, without a vastly increased amount of real-time intelligence information.

In addition, the option of redundancy for all important headquarters could be actively pursued. Redundancy in military equipment already exists to a considerable extent, with stockpiled reserves ready to absorb the inevitable attrition of combat. More consideration needs to be given to the creation of shadow headquarters and installations ready to step into the breach in the event of attack on primary locations. In creating such a shadow organization, it is important to judge the sort of time scale over which the reserve unit is likely to have to exercise its command functions.

It is doubtful if the trained manpower exists to create fully effective reserve headquarters as well as the other paraphernalia for fully integrated C^3I. However, realistically it is unlikely that such units would have to function for any great length of time. While a conventional ballistic attack on a headquarters or other C^3I installations could be debilitating, it is highly unlikely that it could even begin to cause the type of devastation commonly associated with even the smallest nuclear strike. A conventional ballistic missile attack should be thought of more in terms of a conventional bombing mission which these installations need in any case be equipped to absorb. They are therefore likely to be out of action only for the relatively short time necessary to recover from the initial shock of attack and begin repairs on their physical surroundings and the replacement of necessary equipment.

It is, therefore, likely to be the case that such installations may again be capable fairly quickly of resuming at least skeletal military functions. Shadow units would only have to function as fully competent bodies for a strictly limited period of time, while continuing over the longer period to exercise residual authority over areas where the primary headquarters of C^3I installations had suffered more severe damage.

What applies to C^3I headquarters installations also applies equally to other military assets, such as air fields, storage sites or port installations, which are other likely targets of a conventional ballistic missile attack. Again redundancy and duplication offer large possibilities and similar estimations also apply to the level of damage they are likely to sustain, given the amount of conventional munitions targeted upon them. If these installations are unlikely to be rendered totally out of commission, redundancy should be designed not for fully replacing these targets but rather to give a breathing space in terms of a continued availability of some sort of function to military units while repairs are effected.

Such an approach, given the limited number of available conventional ballistic missiles on both sides, would seriously complicate targeting calculations. Everything possible would need to be done to keep such reserve capabilities as secret as possible, while also providing added camouflage by way of decoy sites, increasing the real site's chances of survival by holding out the possibility of a smaller number of missiles per target and also of some targets being totally missed.

In addition, this approach has great appeal to those who favor non-threatening defense in that it contains no aggressive capability and rests purely on passive measures. It therefore has none of the implied aggressive intention which is contained in a policy based on a preemptive first strike on opposition ballistic missile assets. Such an approach therefore would have a great appeal to NATO's political leadership, especially in Western Europe. As stated previously, a military policy based on preemption as the opening gambit of a conventional conflict, which would undoubtedly be presented as a first-strike capability, would be viewed by European publics as a war-initiating strategy. Whatever the threat perception of West European governments with regard to the WTO, it is unlikely to be an approach which could be sold to Western Europe's public opinion.

However, it must be recognized that such constraints have not traditionally been applied to the policy formulation and preferences of the Soviet Union. Defense based on offense has long been the bedrock of the Soviet military's approach to warfighting and preemption has a great appeal to the Soviet Union both in terms of military policy and its political philosophy. In addition, the side which strikes first retains a second-salvo option, which, although blunted by the inevitable NATO response, is still likely to be fairly effective in terms of the residual number of worthwhile targets left within NATO's infrastructure in Western Europe.

10
Conclusion

It seems that the issues surrounding the ATBM debate are now inexorably linked, at least in the minds of politicians and in the press in Europe, with the concerns which surround the implications of the SDI research program and the future of ballistic missile defense. This linkage was mainly due to the coincidence and timing of a more general recognition of the possibility of an emerging threat from conventional ballistic missiles and a decision of President Reagan to tackle head-on the issue of defense against nuclear ballistic missiles. Coupled with this was the early identification of the existence of a high degree of commonality between the technologies necessary to deal with these two problems.

It is, however, in many ways an unfortunate linkage, especially as the ATBM debate has tended to become divorced from the rest of NATO's conventional strategy and the discussions surrounding how NATO should deal with the prospect of a Europe with fewer nuclear systems. The lumping together of ATBMs and SDI has meant that the issue of defense against TBMs has become bound up with the discussions now going on concerning what form of offense/defense relationship is most stabilizing and what the nature of NATO's restructured strategic posture should be. Divorcing ATBM from its cousin, air defense, and linking it to its distant relative, BMD against nuclear missiles, has done it no service. Any technological advantages which may accrue are more than offset by the political problems which will in the future surround any debate on ATBMs.

As this paper shows, NATO is technically capable of developing and deploying ATBMs, at least as point-defense weapons. NATO also has the capability to build the types of conventional ballistic missiles necessary for preemptive strikes on the Soviet TBMs. However, the political realities of West European defense postures mean that as a practical strategy it is a non-starter. Even the limited experience which NATO has had to date with the technologies involved in launch or mid-course interception has shown that the tasks involved in deploying systems which are capable of intercepting TBMs at this point may be beyond, at least for the foreseeable future, NATO's technical and financial resources. Indeed, the fact that there is a continued strident debate within NATO on the actual threat posed by possible Soviet deployment of conventionally armed TBMs in many ways negates the possibility of NATO even considering this possibility. Where there is no generally perceived need, NATO is unwilling to pursue any expensive strategy.

At present the debate on Soviet TBMs seems to suggest that NATO, at least for the next few years, would be faced with inaccurate missiles with low payloads, always assuming that the Soviet Union decides to deploy conventional TBMs. While not seeking to minimize the missile threat, there is little doubt that a far more immediate danger is posed to the same NATO assets by the continued Soviet predominance in tactical air capability. Indeed, the main threat to NATO from TBMs may be as defense-suppression weapons, as a prelude to an air attack. An examination of a possible Patriot upgrade shows that one of the main justifications for a policy of active defense against TBMs may in fact be self-defense for air-defense assets.

If a dedicated ATBM system is to be pursued by NATO, even the limited type suggested by a Patriot upgrade, then analysis suggests that NATO is going to have to find a great deal of cash from very tight defense budgets to pursue what is essentially duplicate air-defense deployments. There seems to be little or no political will to pursue this approach. There is a strong preference to devote what resources are available to far more immediate threats to NATO's well-being rather than on some putative TBM threat whose real danger is at least ten or fifteen years in the future—the likely time scale for the CEPs of Soviet ballistic missiles to begin to approach the accuracy necessary to threaten targets with conventional weapons.

The very confusion which surrounds the whole topic of conventional TBMs and ATBMs means there is little prospect of pursuing any worthwhile arms control measures connected with conventional ballistic missiles. Although the INF Treaty's provisions will impact on the ATBM debate, in that with far fewer short-range ballistic missiles available, there is less chance that any can be spared for a conventional role, it does not seem likely that TBMs will disappear from the European theater. Indeed, the pressure for conventional TBMs may increase as the number of nuclear systems is reduced. It has been reported that within two years the U.S. Army expects to field a short-range ballistic missile that will carry conventional bomblets to destroy tanks and bunkers well behind enemy lines.[29] It is likely that these missiles will owe far more to the multiple-launch rocket systems (MLRS) in their origin than to the Lance missile. Where American technology leads, the Soviet Union usually tries to follow. It would be surprising if the Soviet Union were not engaged in developing a similar type of less sophisticated, but still useful ballistic missile. It therefore follows that a rationale for some form of defense system will continue to exist. These new ballistic missiles may not pose the same level of threat as an improved Lance or SS-21 successor, but they would pose a credible danger to vulnerable assets which will need to be protected. The question then becomes how and at what price? The arguments, therefore, seem set to continue. In this sort of situation such arms control is not likely to find ready supporters in European capitals.

Similarly, the ABM Treaty looms large over much of the debate on the type of ATBM deployments which NATO can realistically consider, especially with regard to the technology-transfer provisions of the treaty. However, it would be wrong to overemphasize the importance of the ABM Treaty to a possible ATBM system. Although SDI-derived radar and sensor technology would be useful, it is not essential for a workable ATBM system designed for point defense. In addition, it must be recognized that should a future U.S. administration decide to adopt the broad interpretation of the ABM Treaty it remains an open question whether the European allies, in spite of what is likely to be vocal opposition from sections of their own publics, would refuse the offer of American technology simply because that technology was held to be outside the restrictive

29. *The New York Times*, January 3, 1988, p. E4.

interpretations of a treaty to which they are not a party. The main impact of arms control may actually be to take the heat out of the ATBM debate. Despite the discussion in this paper of the threat posed by conventional TBMs, for much of the political and public opinion the ATBM issue is concerned with defense against nuclear missiles. As the nuclear missiles disappear, it may be that the ATBM debate will revert to its previous military/technical level. An example of this is that the recent debate in the United States over the early deployment of SDI has taken much of the pressure out of the European ATBM debate and has largely removed it from the pages of the press.

American enthusiasm for active defense to one side, for the foreseeable future it is to passive means of defense that the West Europeans are likely to turn as the threat from TBMs recedes from the public mind. Passive measures have the advantage of being proven and cost-effective in terms of the resources they absorb in return for the protection they offer. If the best that affordable active defense can offer is limited self-defense for the missiles themselves, with some marginal overlap for value targets, then pouring concrete and dispersing these same targets, although less glamorous, may offer more in terms of tangible benefits for NATO's defense posture. Indeed, active defense and conventionally armed TBMs may turn out to be an expensive blind alley. It has not yet been satisfactorily proven that conventional TBMs offer any significant advantages over other means, such as air attack or long-range artillery, for delivering the same weight of munition. When a final threat assessment is actually made of how great an asset conventionally armed TBMs could be, it may be that both sides will decide that they have more pressing calls upon strained military budgets.

ABOUT THE AUTHOR

Ian M. Cuthbertson

From 1981 to 1987, Mr. Cuthbertson was a Senior Research Officer in the Arms Control and Disarmament Research Unit of the Foreign and Commonwealth Office. In that capacity he conducted research on arms control and disarmament issues and assisted in the formulation of official British policy. He wrote this paper while a Resident Fellow at the Institute for East-West Security Studies, 1986–1987. Mr. Cuthbertson studied modern history at the University of Stirling and defense studies at the University of Aberdeen.